Chrystah Cabada

村
原
久
十

Hiroku Matsura

MINO

MINO

A Young Scientist's Life-long Journey

Through Outer and Inner Space

Hisako Matsubara

DANIEL & DANIEL, PUBLISHERS, INC.

MCKINLEYVILLE, CALIFORNIA · 2019

Thanks to NASA and many friends at NASA
for their lasting support.

Copyright © 2019 by Hisako Matsubara
All rights reserved
Printed in the United States of America

The author's royalty earnings from the sale of this book will be donated to the Mino Endowment.

The interior design and the cover design of this book are intended for and limited to the publisher's first print edition of the book and related marketing display purposes. All other use of those designs without the publisher's permission is prohibited.

The names of a few people and places have been changed as a matter of courtesy.

Frontispiece photograph by Tomoko Ishihara.

Published by Daniel and Daniel, Publishers, Inc.
Post Office Box 2790
McKinleyville, CA 95519
www.danielpublishing.com

Distributed by SCB Distributors (800) 729-6423

LIBRARY OF CONGRESS CATALOGING-IN-PUBLICATION DATA
Names: Matsubara, Hisako, author.
Title: Mino : a young scientist's lifelong journey through outer and inner space / by Hisako Matsubara.
Description: McKinleyville, Calif. : Daniel & Daniel Publishers, 2019.
Identifiers: LCCN 2019007315 | ISBN 9781564746146 (cloth binding/hardcover)
Subjects: LCSH: Freund, Minoru M. | Astrophysicists—Biography. | Glioblastoma multiforme—Patients—United States—Biography. | Outer space—Exploration.
Classification: LCC QB460.72.F74 M38 2019 | DDC 813/.54—dc23
LC record available at https://lccn.loc.gov/2019007315

CONTENTS

NASA

` FIGHTING

A section of photos and drawings follows page 200.

This book tells a true story. The author created the narrative in Mino's voice, using thousands of her own dairy entries covering nearly five decades, using Mino's numerous letters home, and using his blog "A Little Detour," written during the final two years of his life.

MINO

PRELUDE

AT THE AGE OF FIVE I must have reported at home that Annette's finch had died, the one with the green wings, yellow belly, and black face. Pastor Zimmerli told us that when we die, we all go to Heaven. "Annette's finch too," he said. "We all go to Dear Lord." But Annette's finch is not in Heaven. Don't you understand, Mama? I tell you, it is not in Heaven. *No, no, no.* It lies there at the bottom of its cage. I saw it. I touched it. Just dead.

This was some forty-five years ago. Now, by the time you read these lines, I'll no longer be around. I don't know where I'll go. Some who want to console me say I'll go to Heaven.

Professor Akert, a deeply religious man who became a neuroscientist, in whose house I lived for my first two university years in Zürich, once said at dinner table: "When you die, you're done. Dead is dead."

A devout Buddhist told me in Japan that Bodhisattvas carry the souls to Nirvana, the Land of Eternal Blessing.

When I sat with Grandpa in the East Room, where we could see the city lights of Kyoto down in the valley, he said that souls live as long as they are remembered. Once souls are forgotten, they slowly fade away, merging back into nature, from which we all came.

I don't want to die. I don't want to go either to Heaven or to Nirvana. I want to stay here, here on Earth, and turn my dreams into reality. I want to bring together the knowledge from the vastly different scientific disciplines I have traveled. I want to reap the fruits of nanotechnology, with its tiniest of tiny dimensions, to cosmology, with its truly unimaginable vastness extending to the edge of the universe. I'm not yet done exploring.

As a graduate student at the ETH in Zürich, I built a scanning tunneling microscope all by myself, one of the very first ever built, in order to see atoms. Atoms—imagine, true atoms. How small are they, really? One millionth of the width of a hair. I was one of the very first humans to see carbon atoms with my scanning tunneling microscope, wiggly carbon atoms of a sheet of graphite arranged in a hexagonal pattern.

Later, when I joined NASA, I became involved in cosmology, the science of the universe, and helped built the detector, with which we could peer deep into space, to near the edge of the universe, billions of light years away. Hard to imagine a light year. When we send a laser pulse to the moon, it reflects and arrives back at Earth within a bit more than two and a half seconds. That's how long the light takes to travel the distance to the moon and back—a bit more than two and a half seconds. A year has more than thirty million seconds. Imagine how far the light can travel in thirty million seconds. Imagine further what it means that a galaxy is five or ten billion light years away, a "five" or a "ten" with nine zeros. The light that I saw arriving from those distant galaxies had started its journey at a time long before there was life on Earth or before the Earth had accreted as a planet out of the disk of dust that surrounded the early sun, or even before the sun existed.

It humbles me to think of these unimaginably large and unimaginably tiny dimensions. It humbles me to know that galaxies form and evolve, that stars are born and die, that everything is changing and evolving. Nothing stays the same— *panta rhei*. The universe, the sun, the Earth and everything on it, continents and oceans, mountains and rivers, all living things on Earth, constantly changing, constantly evolving, constantly transforming through time and space—*panta rhei*.

I don't want to go either to Heaven or to Nirvana. More than forty scientists and engineers are working under me here at the NASA Ames Research Center. All highly dedicated. All full of ideas. All upbeat. It's too early for me to go. I'm at the zenith of my life. I'm not yet done exploring.

I've fought hard to slow down this egregious brain tumor in my right parietal lobe, GBM grade IV. I've fought every day to get rid of the tumor. I've submitted to three major surgeries, to radiation, chemotherapy, integrative medicine, immunotherapy—week after week, month after month, now twenty-nine months since diagnosis.

I close my eyes and try to breathe slowly. Something strange happens to me, as if I were transported into a different state of reality. My breathing deepens. I'm in Big Basin, surrounded by majestic redwood trees. I approach them, watching spots of sunlight here and there. I touch their thick bark, listen to their voices. I feel pulled down, down to the ground, irresistibly pulled to my knees, touching the fallen redwood needles with the palms of my hands.

It's one of those magic moments when all senses come together. I see the sunbeams swirling through the branches. I feel the earth. I hear the sounds. I smell the sweetness of the forest floor. I feel the energy streaming out of the redwood trees into me. An explosion of all senses, just for a flash, just for a moment; and as suddenly as it has come, the vision passes away.

After having fought so hard and for so long, I realize that I am not going to win. I'm losing. I've resigned to the certitude that this tumor spreading into my brain, day by day, hour by hour, will extinguish my spark of life.

Strangely, however, the more I think about it, the calmer

I become. I feel no anger, no regret. I am floating in an otherworldly space, in tranquility.

After I'm gone, friends and others whose paths I've crossed will remember. Saya, my mother, will go through her hundreds, even thousands of diary pages chronicling the five decades of my life. She'll go through my letters and through my blog, which I started after I was diagnosed with this Grade IV GBM. Saya may open her diary pages from years past describing how I watched an earthworm on the forest floor munching on a rotting leaf, how I imagined raccoons dancing in the temple garden under the full moon, how Papa showed me a dandelion flower dipped into liquid nitrogen becoming brittle like glass when I gently hit it with a hammer. She'll be reminded that I was not accepted into elementary school on the first trial because I didn't know the word "obey." She'll smile at the memory of my first piano concert. She'll remember how easily I moved between English, German, French, and Japanese, and how I was at one point infatuated with Voltaire's view of the absurdity of life.

Saya will be going through my black ink drawings. She will be looking at the stone lantern in front of Grandpa's Shinto shrine in Kyoto or at the worshippers in Notre Dame de Paris, which I sketched on an Easter Sunday morning. Their bent-over silhouettes express such utter loneliness and sorrow. Saya will hold in her hands the letters I exchanged with Amelia, my only true love. She'll read how Amelia and I scaled a sheer rock face in the Swiss Alps to reach a hanging valley full of wild gentians, blue as the autumn sky.

Yes, I hear Saya singing by my bedside. She still believes I can win.

EARLY YEARS

MY BABY ELEPHANT

Once I start to walk all by myself, the world looks different. There is no limit to what I can do and discover in our house.

What a difference from the time when I was still learning from Mama how to walk. At that time, she walked backward in front of me, taking my hands and guiding me forward. I gave my full attention to her face. I listened when she said, "Minochan, Minochan, *ichi ni, ichi ni, ichi ni.*" I moved my feet forward, one after the other, one-two, one-two, step by step, *"ichi-ni, ichi-ni, ichi-ni."*

Now, every morning I can't wait to climb out of my bed and run to the kitchen, where Papa sets our breakfast table. Mama usually can't wake up till later, so we men start work around seven o'clock.

At the table I have a high chair, stable and safe. Papa puts me on it and pushes me closer to the table. No matter how much I kick my legs and move my arms, the chair remains rock solid. Because my chair is so high, I can see everything— bread, butter, yogurt, strawberry jam, and sometimes sausage or ham. Papa ties a bib around my neck. I own more than a dozen bibs. He doesn't have any and Mama doesn't either.

When I say *"Aaah,"* Papa gives me porridge, which I eat with my own spoon. *"Ai...ai...ai,"* and he gives me a slice of bread with butter and ham. I have lots of teeth to bite and chew with. He drinks coffee. I drink orange juice. I don't like cow milk...makes me feel miserable. Papa says that's because I drank Mama's milk, only Mama's milk, for my first nine months.

By the time Mama comes to the kitchen, Papa and I have finished our breakfast and he is ready to drive to the institute. I stay on my high chair, so I can see better what Mama does for her breakfast. She cooks rice, which she soaked overnight in a pot. Rice is very precious because it comes from a Japanese store in faraway Hamburg. Actually, Papa tells me that the rice comes from California. It is so expensive that only Mama is allowed to eat it, but she always feeds me anyhow. She scrambles three eggs and pours a few drops of soy sauce, specially shipped from Kyoto and also very precious. I love the sound of scrambling eggs. Raw eggs taste fantastic with a few drops of soy sauce. Mama gives me to drink half of what she has scrambled and pours the rest into a hot pan.

Before leaving for the institute Papa tells me, "Don't tire Mama! Be a good boy."

I never tire Mama. What a suggestion. I always help her— help her wherever I can. As soon as we finish our breakfast, she takes off my bib and puts me down on the floor. I push one of the kitchen chairs to the sink, climb on it, and do the dishes. I clean the bowls, the plates, the spoons, knives and forks, chopsticks too, just like I've seen Mama doing it. I put a lot of care for each item, because everything is larger than my hands and can be slippery. The brush is cumbersome to handle, but I've never dropped anything or broken any cup or plate. While I'm working at the kitchen sink, Mama holds the chair tight on which I stand. Often she catches me when I lose balance.

"*Minochan, Minochan, kirei ne!*" Mama praises me for how well I do the dishes, how good I am at using my hands.

After the long work in the kitchen, we go to my room,

because Mama wants to clean my po. This is fun time. She puts me on a long table and takes my diaper off. I feel so liberated that I must kick my legs, turn on my belly, do push-ups, and start to crawl. *"Minochan, Minochan, kora...kora...."* She catches me, she turns me onto my back again and cleans my po. Before she wraps me with a new diaper, I enjoy once more the freedom to kick my legs into the air.

We have two balconies, one in the front overlooking the Baurat Gerber Street, and one in the back off the kitchen. There Mama hangs our laundry on a row of several clotheslines. She sings. She often sings when we are alone. She sings about the sun and how bright it is in the sky. She sings about flowers in the garden, she sings about children coming home from the playground, and about birds flying across the evening sky. She sings about the moon rising over the mountain. She sings about the stars twinkling at night.

To hang our laundry, I help Mama by pulling out my diapers from the basket, one after the other. I know my diapers well, the Japanese ones with dolls, kites, birds, and flowers on a light blue background; the German ones, huge and just white. At the end, I run up and down between the clotheslines and under the strung-up laundry as the sign of completion of our work.

I have a room of my own with a brand-new bed where all my friends can sleep with me—Teddy, Cat, Frog, Bambi, and the Baby Elephant from Grandpa in Japan. All my clothes are in the closet. I have wooden blocks, red, yellow, blue, and green, which I use to build a tall tower. I have several picture books, lined up on my bookshelf, which Mama put together by pasting wrapping paper on boxes. The floor is covered with a thick rug, where I can crawl.

Nevertheless, I never stay long in my room, because I want to be with Mama. She sits at her desk in the tatami room, reading and writing, with her bookshelf in the corner.

Eight tatami, spread from wall to wall. Woven in Kyoto, they were shipped to Hamburg and brought from there by train to Göttingen, where we live. I like to sit on the tatami and look at the scroll hanging on the wall. Mama also likes to look at the scroll, in silence, with eyes full of happiness. I do the same. Calligraphy from Grandpa.

On the window side, we have heavy curtains from the ceiling to the floor, lots of red on white. There I hide, and Mama has to look for me. One wall of the tatami room is a huge mirror. There I can see myself dancing with Mama to the music from our record player. I can see me reading from my picture book to her. I can see me drawing on a large paper spread out over the tatami. I can see me playing with my friends that I bring over from my bed.

One Sunday afternoon Papa got the idea of taking a picture of me sitting on the tatami with my baby elephant, his ears flapping high. While Papa was busy adjusting his camera, Mama put a cotton kimono on me and I sat on my heels like she always does. I was one year and two months old.

EARTHWORM

Papa takes me to the forest. I run through the dry leaves... *kasha...kasha...kasha...*run, fall, stand up by myself, and fall again. Once I see something moving between the leaves. Papa tells me it is an earthworm. He picks it up gently and shows me how it moves. He says it is eating the soil, working hard in the summer and sleeping in the winter deep below ground. I carry the earthworm carefully back to where I had found it. Does the soil taste good, I wonder? Next time I fall down, I put my finger in my mouth and lick it to see how soil tastes. No taste. I feel sorry for the earthworm.

I see a lot of color flying in the meadow. I run after them, and Papa says they are butterflies. I want to catch them, but they are faster. They fly from flower to flower. I look at the flowers. Papa says they are dandelions.

Pointing at the flower he repeats, "Dandelion."

I point at many things in the forest. Papa says "Tree...branches...leaves...flowers...dandelion...bees...butter-flies...." I point at everything, again and again, until Papa gets exhausted and carries me back home on his shoulders.

At home we have lunch, but Mama is not there. She left for a faraway clinic by train. She had gotten pain, horrible pain in her neck, shoulders, and arms, so that she could no longer lift me up. Papa says the winter was too harsh for her and she didn't eat enough.

But it is summer now. I want Mama to be here with us.

Even before Mama left, I had felt something was different with her. I followed her every step, wherever she went, and stayed close to her. Even when she went to the bathroom. As often as possible, when she sang, I put my head in her arms.

After lunch Papa takes me to the nearby park, where I can play in a sandbox, pour sand over my head, slide down a slide a dozen times. I run with Papa across the lawn to see who is faster. He always loses the race. Coming home, I go from room to room, but Mama is not there.

Papa also looks sad. So, I don't cry. We quietly eat the dinner Mama cooked before leaving. We both feel lonely and want to go to bed.

But first Papa has to change my diaper. He pushes me down on the long table without a word, puts my diaper on, puts on my pajamas, and is done. I'm disappointed. No time to kick my legs in the air, to turn onto my belly, to do push-ups, and to crawl over the long table. Mama would say, *"Minochan, Minochan...kora...kora!"* and I crawl faster until she catches me. It's such fun. But Papa packs me up in a minute and ruins all the fun.

I'm angry at him and start to cry. When I hear my crying voice, all the sadness I've endured since we came home pours out. I cry as loud as I can.

Papa gets mad at me. "What's the matter? Be quiet," he says, staring at me with an angry face. "You have eaten, and I have cleaned you nicely. Stop crying!"

I'm determined to cry. I howl and choke up. I'm sad, very sad, miserable. *Mama is gone...Mama is gone....*

Papa barks at me, "What's wrong? What do you want? Stop crying!" He has no idea how I feel. He doesn't get it. He just doesn't get it. I cry as loud as I can. Out of exhaustion I fall asleep. In the morning, I find myself in Mama's bed, next to Papa.

After breakfast Papa takes me to his institute. I always heard that he must go to the institute every morning, work there all day, and come home for dinner. Now I can see by myself what the institute is. We drive a short distance, park the car and walk to a tall old building. Several men are going in and out of the entrance. Papa carries me in his arms climbing up the long stairs and enters a room. "My office," I hear him say. Two high windows looking out into trees.

Many books in the shelves. More than I have in my bookcase, but much fewer than Mama has. A few chairs and one big desk. Lots of drawers. Papa shows me what's in the drawers. Papers, papers, only papers. No toys. Not a single picture book. No music from a record player.

Papa spreads a large sheet of paper on the floor, takes out crayons from his pocket, and tells me I can draw as much as I want. He must now go to see his students in the lab.

When Papa comes back, he finds me sleeping on the floor. I have drawn many things—cat, dog, frog, my teddy, and my baby elephant. Mama would say how well I have drawn everything, but Papa can't figure out what I have drawn. He only sees circles, circles, lines and is more concerned that I may have caught a cold sleeping on the floor. He carries me to the car and drives me home. There he makes a hot bath for me. After lunch he seems to have decided he can leave me alone in my room. Now I know what the institute looks like. Not such an exciting place. I promise Papa I'll not cry.

As long as Mama is away, I stay alone in my room, twice a week for a few hours. Sometimes I sit behind the entrance door, listen to the footsteps in the staircase, calling out "Papa!" but no answer.

One evening Mama called from the clinic far away. Long-distance calls are very expensive. We had to make it short. Mama said "Minochan, Minochan, *iiko ne.*" I nod at each word, thinking of Mama in my heart.

MY WHITE HAT

During our meals Papa neither speaks nor smiles. He chews, swallows, and takes the next bite. Between bites he looks somber. He never tells stories at table. I'm more used to Mama, who listens to the birds singing outside our window, who says how delicious the soup is, how beautiful the evening sky. Mama smiles and laughs.

Papa looks at me and says, "Eat!" When I tap on the table with both my hands, *tataratata,* Papa says, "Eat."

I'm bored. No fun. I let my hands tap more on the table… *tataratata…* Papa gets angry. "Eat! You still have things on your plate and now you're dropping them all on the floor."

I'm proud of eating all by myself with my spoon. I may drop a few, but I'm doing my best. Papa says, "Eat! I've cooked a good dinner for you. Eat it now. It's getting cold. Eat."

Mama would start singing a song we have been practicing in the tatami room, *"Chi chi pappa chi pappa…"* The song is about little baby birds learning to sing under a teacher bird. I join Mama and we both tap the table…*tataratata…*and sing: *"Chi chi pappa chi pappa…"* That's how I get hungry and eat more.

While I was asleep at night with my friends in my bed, Mama must have come home. In the morning, in the kitchen,

Papa tells me the news. I run to the bedroom, see her sleeping, and call, "Mama!" I jump into her bed. She wakes up and takes me tight in her arms. "Minochan, Minochan, *arigato ne!*" She starts to cry. Papa comes and puts his arms around her. We all three cry and laugh together.

After breakfast Papa leaves for the institute. I take Mama by the hand and show her in our tatami room what I can do. I can kick my left leg so high that it reaches the edge of her chair. I do the same with my right leg. Mama is impressed. "Minochan, Minochan, *sugoi ne!*" I spread my legs wide, put my hands on the floor, and bend my head down. I can see Mama upside down. Everything else in the room is also upside down. Mama bursts into laughter. I'm happy she laughs.

We are going to town. I now can walk from our condo down the stairs to the first floor, holding Mama's hand. She doesn't have to carry me down anymore. She puts me in my little carriage and off we go. She has a list of things to buy.

We drop into a children's store. A shop girl tries several pairs of shoes on my feet, very wide with a high instep. I walk around. Together with Mama I choose a pair of black shoes with black laces. Mama picks up a white hat and puts it on my head. She shows it to me in a mirror. I like the hat and want to keep it...keep it on. Mama talks to the shop girl and buys in addition some roundish thing which the girl wraps in paper.

We go to the butcher, where I usually get a slice of my favorite ham. We go to a vegetable store. According to Mama, in Germany, they don't have many vegetables and fruits to choose from, but she buys a few. At the end of our trip through town we go to the pony in front of a toy store, where we always go. Mama puts me on the pony and lets some coins fall into the slot. I ride and ride the pony until it stops.

On our way home through narrow lanes I ask Mama, pointing at many things I see. She answers. "A child...a bicycle...a post office...a restaurant...a man...a bookstore..." and many more words. This fascinates me because everything seems to have a name.

At home, Mama unwraps the roundish thing she bought. A blue pot. Mama tells me to sit on the pot and do pee. Holding me with both her hands, she says, "Peeee!"

I say, "Peeee!"

She repeats the word, but I have no idea what she wants me to do. So I sit there on the blue pot, wearing my new white hat and a red shirt, being held by Mama, who says, "Peeee!"; but I can't figure it out.

After several attempts, Mama gives up and packs me again in a diaper. Every day she tries out the same. If only I knew what she wanted. One day I don't see the blue pot anymore. Mama changes my diapers as before. I can have fun again, kicking my legs, turning over, doing push-ups, and crawling around, until Mama catches me.

By the way, I love a hot bath. My frog and my duck join me. I dive with them, come up again, dive in and out and in and out. Mama cleans me with a wash towel and I do the same for my friends, the frog and the duck. Bathing takes a long time. Mama sits on a chair in the bathroom, reading a book or writing in her notebook. If Papa is at home, he and I bathe together. I splash and spray him as much as I can. We let the frog and the duck swim between us. We play until we are exhausted.

Papa likes my new white hat. "It protects his face from the sun because of its broad brim," he says to Mama.

Mama smiles, "Oh yes, it's made just for him."

I keep my hat on day and night, except for bathing. I sleep, eat, clean dishes, hang diapers, draw and read, sing and dance, go to town—always with my white hat on my head.

MOO'S VISIT

It's getting chilly outside. I like to stand on the balcony next to our living room and look over the garden below. My hands on the rail and my face just above it, I can see the leaves on the trees. Yellow and brown leaves, some even red. *"Happa,"* I tell Mama, who squats down next to me, her hands also on the rail. "Summer is over. It's fall already. *Happa* are falling." Yes, *happa* are tumbling through the air, falling down to earth. Mama adjusts the brim of my hat, so that I can see better.

A man is walking past with a dog.

"Wow wow!" The man turns his head up and waves to me. I wave back.

A woman walks on the other side carrying two bags. She doesn't hear me when I try to greet her. She walks faster and is gone.

In the condo downstairs a girl lives with her Papa and Mama. Upstairs there is also a girl. Both are about my age. I visited them once. Both have many toys, many more than I do. Their rooms are really filled with toys, but they don't play with them. They are watching TV. Their mamas also watch TV. The TV is always on.

We don't have a TV, but we do have a record player. Mama has many records. She calls them LPs. I already know how to put an LP on the record player, but to place the needle, while the LP is turning, is difficult. Papa tells me I can do it when I am five years old. I'm now one year and ten months. So very soon.

I dance with Mama, when the LP plays music. I know several pieces like *"Sho sho shojoji..."* It's about a temple garden filled with moonlight. Raccoons come out and dance, tapping their bellies. I have a drum from my uncle in Japan. I tap it, because I'm the biggest raccoon. Mama and I play the music and dance again and again. I never get tired, but

Mama gives up. We eat, take a nap, and then go out to the green park or into town.

Once Papa takes us by car to a nearby village. There I see a large creature. Huge black and white body. A cow. I never imagined how big a cow really is. She's chasing flies with a long rope at the end of her body. "That's her *shippo*," Mama says. Moo has a *shippo*, a long *shippo*, chasing flies. Papa lifts me up and pats the moo. I pat her too. At this moment she makes "moo…moo" and shakes her head. Flies bother her around the eyes.

At night, Moo comes into my bedroom chasing flies with her *shippo*.

"Moo! Moo!"

Papa comes running. "What's the matter? You just had a dream. Now go back to sleep."

I'm wide awake. "Moo! Moo!" I try to convince Papa.

"Listen! There is no moo here. Go back to sleep."

The moo *was* here. Papa doesn't get it.

I want apple juice. Papa brings apple juice. "Drink and sleep," he says sternly.

I start to cry.

Mama comes with sleepy eyes. "Minochan, Minochan, *nenne yo*." Papa tells her about moo. She takes me in her arms. "Ah, yes, Moo came to see you. Ah, yes, Moo was here chasing flies with her *shippo*, but now went back to the farmhouse. Moo *nenne*. Minochan *nenne*."

I feel better and get sleepy again.

Since Moo's visit in my room I have gotten into the habit of waking up after midnight. I cry until Papa or Mama, or sometimes both, come to me and do something. The best would be to play music and dance with Mama, "*Sho sho shojoji*," I say, tapping my drum. Both look sleepy. They are not listening to what I say. They are just collapsing. Mama adjusts my white hat and Papa brings a cup of apple juice. At the end they take me to their bed and let me sleep between them.

Mama reads Dr. Spock's book. She finds a passage about what to do when a baby cries during the night and sleeps during the day. Mama talks with Papa, wondering if this applies to me. They aren't sure. I'm not sleeping all day long.

On another page, Dr. Spock's book says that parents should not come to the crying baby but let him cry till he stops and falls asleep.

Upstairs and downstairs bang the ceiling and the floor, whenever I start crying at night. The woman upstairs tells Mama to give me a sleeping pill. "In our country you can buy sleeping pills for babies," she must have lectured Mama.

The girls upstairs and downstairs never cry during the night, because they are good babies. Or they get sleeping pills. Upstairs is the family of a bank director and downstairs the family of the director of the largest department store in town. Both women stare at us when Mama and I go out or come back.

Dr. Lezius, our family doctor, comes to see me just as I'm using a screw driver to open Papa's alarm clock to see what's inside and how it ticks.

"So, so," he says, "you're working hard, Mino." He checks me out, looks into my nose, ears, mouth, and eyes; listens to my heart; and tells Mama that I'm all right, I'm not sick. Nothing wrong with me.

"Soon he's going to sleep all through night," Dr. Lezius promises. He pulls out his pocket watch with a long golden chain. It has a cover that flips open.

"Should be going," Dr. Lezius says, but before leaving he lets me hold his pocket watch in my hands. I flip the cover open and close again. Papa doesn't have such a nice watch.

MY DUCK IN HEIDELBERG

Papa must travel to Heidelberg. Mama packs his suitcase and I help her. To a conference, for three days, Mama tells me, packing his suitcase. I carry Papa's shoes, boots, and sandals one by one and put them into his suitcase.

"Papa wears the same shoes three days and doesn't need others," Mama tells me. So I take them back to the shoe cabinet one by one.

I bring my duck, because Papa likes my duck when I'm with him in the bathtub.

Mama tells me how happy Papa will be with my duck in Heidelberg and packs it carefully between the clothes. "Minochan, Minochan, *arigato ne*!" she thanks me for my help.

We three have lunch together at home, but then Papa must leave. I go with Mama to the balcony and see him off. "Papa!" I call. He waves back from his car. He will be happy with my duck in Heidelberg.

For the next two nights, I sleep in Papa's bed.

Mama tells me a story of a good boy, who has seen earthworms, butterflies, and dandelions in the forest, who likes to watch yellow and brown *happa* falling in the garden, who can kick his feet high in the air, and can look at the world upside-down. This boy sleeps through the night. All his friends are already asleep, and Moo sleeps too.

Then Mama sings *"Nemure yoiko yo…"* I know the song so well. I've heard it many times and want to hear it every night. Two more songs will follow, *"Nemure nemure haha no muneni"* and *"Nennen yo okororiyo."* All songs are about sleeping children. I like them, but I don't want to sleep yet. I want to hear more songs and once more the story of a good boy.

Mama tells me, if a boy doesn't sleep, he can't grow. If he stays small, he can't place the needle on the record player.

She sings two more songs, one about the crows flying back to the forest for a good night's sleep and the other about the stars high up in the sky watching over sleeping babies.

I still don't want to sleep. I want to be with Mama. I know when I fall asleep, she goes away. I kick my legs and want to drink apple juice. She brings apple juice and covers me better. I look for some other reason not to sleep. Finally, Mama gives up, and we both fall asleep.

Three days have quickly gone by. Papa comes back from Heidelberg. He tells me he was happy with my duck there.

Mama is practicing a speech in front of our mirror.

Before leaving for the institute Papa tells me: "This afternoon Mama is going to give a talk in German. Don't disturb her. Be a good boy!"

I am a good boy. I sit on the tatami and watch Mama talking to the mirror. She talks and talks and talks. It's boring. She should dance instead. She should sing.

"Mama!" I start to sing *"Sho sho shojoji..."* and tap my drum.

Mama bursts into laughter: "Minochan, Minochan, *dame yo!*" She tries once more to talk to the mirror. Turning her head left and right, she makes serious faces all the time, which she never does at home.

I get bored, stand up, and dance *"sho sho shojoji..."* Lots of raccoons come into the temple garden in the full moon night, *"Sho sho shojoji...Mama!"* I call her, because now it is her turn to sing. She sings, and I tap my drum. We both dance *"Ponpoko pon no pon."*

She puts my new red sweater on me, the one we had bought together in town, and my dark-blue pants. I keep my white hat on. Mama wears one of her kimonos, lots of white chrysanthemum on red, the same red as my sweater.

Papa takes us by car to town and we all enter a big building. Mama disappears somewhere. "She must go backstage,"

Papa tells me. He carries me into a big hall. Many men and women are sitting in long rows. No children.

Papa carries me all the way to the front row. There Dr. Lezius, our family doctor, shakes hands with Papa and smiles at me. Papa takes me backstage, where Mama and a man in a blue overall are testing out something that Papa calls a microphone.

I walk around and find a long curtain hanging to the floor like in our tatami room. I pull the curtain aside and see many faces looking up to me. They start clapping. Dr. Lezius comes running to the edge of the stage and reaches up. He carries me back to the seat. The lights go out. The curtain opens. Mama comes onto the stage and starts to speak with a serious face. I like her in her kimono. Her voice echoes through the microphone. It sounds much louder than her voice at home. She talks and talks, and her voice is always loud.

I'm getting interested in the microphone. I want to see what's in it. I want to pull it apart. Mama's voice still echoes.

After her speech Mama starts to dance with the music from her record. I know that music and dance. At the end everybody claps and claps and claps for a long time.

I clap too. "Mama!" I call as loud as I can. She turns toward me and smiles.

KARIUS AND BAKTUS

Today is my birthday. I've received many presents from Marburg, Wiesbaden, Switzerland, Japan, and America. Mama has written down what came from whom. I'm happy about a picture book with many stories. It's from Mama. I have already looked through it. A very heavy book. She is going to read from it for me every night. My bookshelf is filling up with new books. I know them all and where they are. My toys are all over the house, but my books are all neatly in the bookshelf.

Papa gives me a present too. He calls it a "magnifying glass." He shows me how large my hand becomes when I look at it through the glass. He says we can look at all kinds of things in the forest when spring comes. I'm really curious.

Mama measures my height and weight and reports to Papa in the evening how fast I've grown. She shows him the page of her notebook where she has drawn my right foot with a black crayon. I remember how ticklish it felt to stand on the page without moving, while she went down on her knees and drew with the crayon exactly along my bare foot. She needed only one foot and compared it to the largest potato we have in the kitchen. "Much larger!"

Next day she makes the drawings of my right foot on many papers to send them out to all who had given me birthday presents. Now everybody knows how large my foot is—larger than the largest potato in our kitchen.

Papa takes me to a park. Snow, snow, everywhere. I'm wearing a warm red cap. It's from Oma in Marburg. My white hat has become too small, because Mama washed it too hot. I gave it to my Teddy, but he didn't like it because he can't see a thing with this hat on his face. It is now on the head of my baby elephant, who supports it with his wide ears.

I look for *happa*. Papa says they are buried under the snow. I dig and dig. My hands get very cold. Finally, I find a few brown *happa* and pull them out. They look tired. Papa says, they want to become soil. *Happa* become soil? Everything that lives becomes soil. I put the brown *happa* back, covering them with snow again.

On a hillside, I ride a sled with Papa. I sit in front and Papa in the back. We sled fast. Then we climb back up to the top of the hill, stomping through the snow. We sled downhill again, fast, really fast. We both like it so much that we don't stop until it's getting dark. Papa pulls the sled with me on it and walks to our car. The snow is so white that it doesn't get really dark.

At night, Mama reads out of my new picture book. I sit in my bed with all my friends and listen to her. She shows me the page with two small boys, one with red hair and the other blond. They are Karius and Baktus. They are so tiny that we can see them only under a strong magnifying glass.

Karius and Baktus work together every night in the mouth of a boy while he is sleeping. They go from tooth to tooth, looking for what the boy has eaten for dinner. They look between the teeth and behind the teeth. If they find bits of food, they become strong. Then they start digging to make a hole. They sit in the hole, chat and dig deeper. After some time, the boy gets a toothache and can't eat anymore because it hurts.

Listening to Mama, I open my mouth and close it. I wonder if Karius and Baktus are working in my mouth going from tooth to tooth. I wonder if I can hear them chat. What are they talking about?

Mama tells me Karius and Baktus can't find anything in my mouth because my teeth are clean. She brushes them every night, though I don't like it. She tells me Karius and Baktus can't find any food in my mouth. In the morning, I'm not going to have a toothache.

I feel relieved but also disappointed because I can't see Karius and Baktus working in my mouth, even though I now have a magnifying glass. Nor can I hear them chatting. How do they dig into a tooth? Do they carry a screwdriver? I found a screwdriver in one of Papa's boxes long ago and used it to open his alarm clock. Or do Karius and Baktus dig in a tooth with a shovel like Papa does with the snow? Do they carry a shovel everywhere in a mouth? If they don't find any food between the teeth and become hungry, do they cry? Can I hear them cry? I don't want to have any pain in my teeth, but I don't want them to be crying for hunger either.

Mama finishes the story and starts singing *"Nemure yoiko yo"* but I'm still thinking about Karius and Baktus.

Why do they work during the night? Only during the night?

Mama sings the next song, *"Nemure nemure haha no muneni."* And then the last one, *"Nennen yo okorori yo."*

HIDE 'N' SEEK

Mama's theater people come. They take off shoes at the door. They sit on their heels or cross-legged in the tatami room. They are actors from the Deutsches Theater not far from our house, where Mama often goes. She helps put plays on stage. Papa says some of the actors are really famous, like Günter. I sing *"Sho sho shojoji"* in front of them. They didn't know that raccoons come out into the temple garden under the full moon. They didn't know that the raccoons dance in the temple garden under the full moon. At the end they clap. Günter lifts me high up.

The days are getting warmer. Papa takes me to the edge of the forest and shows me under the magnifying glass a bud on a twig just opening. A few tiny *happa*, baby *happa* are coming out bright green. Papa shows me brown *happa* on the ground. They are from last year. We dig and dig and find a tired-looking, old *happa*, which has lost its shape. Old *happa* become soil. All *happa* lying on the ground will someday become soil. "They return to soil," Papa says.

One *happa* is moving, I pick it up. Under it an earthworm, eating the *happa*. Under my magnifying glass, the earthworm looks blind and fearless. We watch how the earthworm is munching the *happa* bit by bit.

At home, I play hide and seek with Mama. She has to stand in a corner, close her eyes, and count to ten. Then she starts looking for me.

"Minochan, Minochan, where are you?"

"Here!"

"Aha, under the kitchen table..."

"No!"

"Oh, I know, under Papa's desk..."

"No! I'm here!"

"Haaa, Mino's voice comes from there...from behind the long curtain..."

I'm very still, trying not to move. Mama touches the curtain, comes very close but gives up saying, "Minochan, Minochan, where are you?"

I come out from behind the curtain. She is stunned. I jump into her arms.

We continue the game. I hide in Papa's bed, in the bathtub, in Mama's kimono-closet, or behind the sofa. If she finds me, I stand in the corner, close my eyes and count one, two, three and hundred. I can count one, two, three and hundred. Sometimes I count one, two, three and million. When I feel like it, I count million, million and hundred.

I find Mama most of the time. Only once I couldn't. Nowhere. She was gone. I became frightened and called "Mama!" I started to cry. She was hiding in my bed, covering herself with my blanket. Next time, I hid in my bed, covered myself really well with my blanket. Mama couldn't find me.

"Minochan, Minochan, where are you?"

We also play fairy tales in the tatami room. One tale I like very much is Urashima. I'm Urashima on the beach chasing the bad boys away. They were throwing stones at a Mother Turtle who had crawled onshore. After the bad boys leave, Urashima pats the Mother Turtle on her back and watches her swimming away. One year later the Mother Turtle comes back to the same beach, where Urashima is waiting. She carries him on her back into the ocean, deep down to her coral palace. Urashima makes friends with many fish and other sea creatures, listens to music, eats delicious food, until Mother Turtle carries him back to the beach. Mama is the turtle, swimming all over in the tatami room carrying me, Urashima, on her back.

I shave my face with Papa's electric shaver, working on it for a long time. It makes *booon booon booon*. I often watch

Papa using it and want to try it out on myself. The problem is how to reach the shelf where he puts the shaver. One day he left it next to the faucet. I knew how to switch it on. It starts *booon booon booon*. I go over my cheeks, nose, chin and forehead many times.

Afterwards, I reach Mama's shelf, take her lotion, and rub it over my face. I scoop her cream and smear into my hair, ears, hands, and pants. Mama has three lipsticks, bright red, dark red, and pink. I use them all around my mouth and draw many lines and circles on both my cheeks.

I look for Mama. She is reading a book in the tatami room. Approaching her, I call "Mama!" She turns around, sees my face and starts to laugh so hard that she falls down on the tatami. After a while we both go closer to the mirror, and I look at myself for the first time from head to toe. I must say I'm quite a mess. My eyebrows are half-gone, particularly the right side. My hair is rising up in the center, supported by a lot of cream. My cheeks, nose, and mouth show many red dots, as if I had spilled a spoonful of strawberry jam.

Mama has an idea. She brings her theater make-up case, cleans my face and draws long whiskers like the cat in the Bremen Town Musicians, who scare the robbers away. She uses her brown pencil to restore my eyebrows and combs my hair taking a lot of cream away. We look in the mirror and at once I feel much better.

Now we play the Bremen Town Musicians, since I'm a cat with whiskers. I'm so excited to play the cat that I forget to be the dog and the rooster at the same time. Mama as the pony is on her knees, singing *eee-aaa, eee-aaa, eee-aaa*, and I'm the cat, crying *meow-meow-meow*. I look at my whiskers in the mirror and sing again and again. "The robbers are fleeing!" I tell the pony. "They are gone!" Mama stands up.

We go to the kitchen. Mama starts peeling potatoes. I want to help her. She gives me a knife and a potato. She shows me how to hold the knife, because it is so sharp, and how to peel a potato. It is difficult to hold the potato with one hand and to work on it with the knife in the other hand. I chip off pieces of potato skin, careful not to cut myself. I chip off more and more, looking only at my potato. Finally, I'm done, and the potato has become as small as my thumb.

But I'm proud and happy. I haven't cut myself. Mama is very happy too and breathes deeply. She's all sweaty. She cooks my potato together with the rest and shows Papa at the dinner what I have achieved all by myself. He is impressed. I give my potato to Mama to eat. "Minochan, Minochan, *arigato ne!*" Thanking me, she chews it for a long time.

After dinner, I go to my room and play with blocks. A few days ago, I was trying to build a tower. I put blocks on blocks, but at a certain height they always fell down. I was mad at them. Papa came in and saw the problem. He shows me how to put two blocks first and then one block as a bridge, continue with another two blocks and one block again as a bridge. I get my tower to the height of five stories, though at the end it collapsed, because I pushed the last bridge too hard.

I started all over and built a tower seven stories high, higher than the one in the book. I have no blocks left in the box, and Papa tells me he would bring more next time he goes downtown. He shows me how to build fences with the skinnier blocks and a farmhouse where many moos can live. I want to build everything all by myself and I do it. Papa is impressed

MAESTRO

Mama's friends at the Deutsches Theater have given her two tickets to a concert, but Papa is attending a conference in Munich. I take his place. Mama dresses me with a new light-blue summer suit, white socks, and black shoes. She's wearing a dress with blue and yellow flowers.

It's warm and the evening sky is turning red. Mama points at the sky and sings *"Yuyake koyake."* I sing along and we

walk through narrow passages all the way to the Concert
Hall.

It is a big building. Many people are going in. All look
very serious. We take our seats on the first row, just in the
middle. I climb up on my chair and look around. I have
never seen so many people together. Some are talking, others
are reading the program, but no children. Mama whispers I
should sit down and wait for the concert to begin.

At home she told me that there will be many musicians on
the stage. I'm going to see violins, cellos, flutes, drums and a
piano. I can hardly wait for the curtain to open.

I ask Mama, whether they sing *"Chi chi pappa"* or *"Sho
sho shojoji."* She says, no, they are going to make music like
the music on our records.

I ask her if they know Bremen Town Musicians, the pony,
the dog, the cat, and the rooster. She says, "Oh yes, they know
them, but tonight they are going to play different music."

Finally, the curtain opens. Many men and a few women
are sitting on chairs. They stand up when an older man climbs
onto a box on the stage. He turns his back to us and raises
a long chopstick. Everybody starts to play. The man waves
his chopstick up and down, right and left. He moves as if he
wants to dance, but he doesn't dance. I climb on my chair,
move my arms like him, but Mama catches me just before I
lose balance. I sit down on my chair and move my arms up
and down, using my finger as the chopstick.

Music goes on and on, and the man bends his head
forward, backward, turns his body sideways, stretches his
chopstick toward the back, up and down, right and left, and
still continues on and on. Finally, he stops. The music stops.
He turns around and bows. Everyone claps. He bows again
and again. His hair falls over his face, but he throws it back
with one hand.

The next music starts. The man does the same. I feel
sleepy and crawl onto Mama's lap. I'm listening to the music,

but it's fading away, far away. I feel Mama's hand holding me tight. After the concert she must have carried me back home, through the narrow lanes, all the way to our house, because in the morning, I find myself lying in Papa's bed.

Mama is sleeping next to me. I don't wake her up. I go to the kitchen and start setting up our breakfast table. I go out to drive with my tricycle up the Baurat Gerber Street, around the corner to Frau Fichtner's store. She has big hands and a big nose. She gives me two rolls for Mama and three for me. She puts them in a bag and hangs it over the handle of my tricycle. Back home I realize I can't open the door to our condo, and I don't have the key. I try to reach the bell but it's still a bit too high. I knock and knock. Mama comes running to the door and we have a wonderful breakfast.

I bring all my friends to the tatami room. Teddy, the dog, the cat, the Bambi, the baby elephant, the frog, and the duck. I stand in front of them and wave one of Mama's chopsticks. Up and down, up and down, right and left, right and left. I sing:

Sho sho shojoji
Shojoji no niwa wa
Tsun tsun tsukiyo da
Mina dete koi koi koi
Oira no tomodacha
Ponpoko pon no pon

At the end, I turn around and bow many times. Mama stands in the door and claps and claps and claps.

URASHIMA

After lunch, Mama goes back to her desk and writes something. I decide to call Papa in the institute. I know the number and how to dial. I hear a man's voice that I've never heard before. This is not Papa.

"Hello, hello," I say.

"Police here. How can I help you?" the man replies.

"Here is Mino," I say, "Papa is in the institute. He does experiments. They sometimes explode—*bang...bang...bang...*

"No, not here. In Papa's lab. Lots of experiments. The kids sometimes make mistakes. No, no, big kids, students, yes, in Papa's lab...

"I'm Urashima and Mama is Mother Turtle. She swims and swims in our tatami room. Lots of water...lots of waves... no, no, we are not flooded...

"Mother Turtle swims to the bottom of the ocean to her coral palace...

"No, no, no, here in our tatami room. Mother Turtle invites Urashima to her palace. Oh yes, Mother Turtle swims and swims in the tatami room, full of water, lots of fish...

"Oh, no, no, blue water and the coral palace at the bottom of the ocean...

"Oh yes, Mother Turtle swims. I know it. I'm sitting on her back...

"No, no, our tatami room not flooded. It is dry...you don't need to come..."

I hang up.

O boy...o boy...o boy...o boy...*o boy!*

PETER AND THE WOLF

I go to the park with Mama. She carries everything in a bag—my red scoop, my yellow sieve, and my blue bucket. Some kids are playing in a big sandbox. I'm going to build a castle like in my fairy tales. With my scoop I sit in the corner of the sandbox. Suddenly one boy throws sand over my head. I ignore him and continue building my sand castle. I gather a lot of sand. While I fortify the center of my castle, another boy comes and kicks all I have built. I start to cry, and Mama says we'd better leave the sandbox and go to one of the slides. Last time she scolded such a nasty boy, but the boy's mother, with a ferocious face, barked at Mama. Today we ignore them. We leave for the slide.

No kids are around the slide. I climb up, slide down, and climb up, and again slide down. It's such fun. I could go on all day. The slide is only for kids. Mama stands at the bottom and watches me coming down. She's laughing. After a while she starts to look bored. It's also getting cold. It's autumn and the sun is weak. Mama wants to go home. I say, "After three more slides." She nods and waits. Then we walk home. I can tell Mama is tired, because she doesn't sing. So I sing *"Yuyake koyake."* The evening red is spreading before our eyes. She smiles and joins in singing.

Papa brings good news. Starting next week, I can go to a kindergarten in a village on the other side of the hills. Papa can drive me there in the morning and pick me up in the afternoon. He says all children play games, sing songs, draw with crayons, and listen to many fairy tales. I can hardly wait.

Next week comes. I'm introduced to all the kids. I'm the youngest, not yet three. Others are between three and six. They are noisy and beat each other up whenever they can. They toss things around, hurl books, crayons, and whatever else they can pick up. Auntie Gertrud, Auntie Marie, and Auntie Ingrid are teachers who give lunch boxes to all of us. While we're eating lunch, a piece of bread, apple, or cracker flies from somewhere.

I'm disappointed. We don't sing together. We don't play games or listen to music or to fairy tales. Sometimes we are all asked to draw with crayons. Auntie Ingrid pins them to the walls. We run around in the large playground, throwing balls, making noise, and beating each other up. There is no slide, no swing, no sandbox.

Dittmar is the oldest. He is big and strong. He walks around in leather shorts from chest to knee. Everyone is afraid of Dittmar, because he can flatten you to the floor with one punch. "He is retarded," some of the girls tell me. "Dumb, cuckoo. His father drinks and beats him up all the time."

I feel sorry for Dittmar. He comes to me and protects me whenever other boys try to beat me up. He never beats me up. He stands beside me, putting his arm around me, when the others come. He is so big and strong that nobody can hurt me. Dittmar doesn't talk much. He only says "damn it" to everything. He doesn't read books. He doesn't like crayons. He loves football. He can kick the ball farther than any other boy.

Today Papa, Mama, and I walk over the hill to my kindergarten. It's a long walk. Spring is here. Lots of green *happa*. We follow the trail up the hill and down. Papa takes my hand and we step off the trail. We stamp through the forest. I like how the dry leaves rustle when I kick them with my feet... *kasha...kasha...kasha*. Papa shows me a thin tree with thin dry branches. "This one couldn't make it," he says. "It didn't get enough sunlight to stay alive." He wiggles it. "This tree is dead. You can knock it over."

I feel sorry for the tree and don't want to knock it over. Papa says, "Fungus and insects will eat it up in the coming months."

"If I knock it over, what'll happen?"

"The tree will lie down and slowly become soil."

I push it with all my strength. After many attempts, I knock it over.

Papa says, "Now the tree rests and little by little will turn into soil."

"Like many brown *happa* from last year?" I make sure. He nods. Pointing at the roots of the dead tree, he tells me, "You see, they are already eaten by fungus. Now you have knocked it over, the tree can rest. You are a strong boy."

Mama is looking around, singing, "Oh, what a beautiful

morning, oh what a beautiful day, I've got a beautiful feeling, everything's going my way..." She jumps around. She
breathes deeply. "The air is very tasty." We walk, run, and
walk for more than two hours and finally arrive at the
kindergarten.

Next afternoon, we go to the Concert Hall to see and
hear *Peter and the Wolf* by Prokofiev. I know the music from
one of our LPs, but I have never seen any wolf. We sit again
in the first row, center, where I sat with Mama last time to
watch the man with his chopstick. Now there are children.
They are talking and running around. Mama tells me, Günter,
who came to see us at home a few times and listened to my
singing *"Sho sho shojoji,"* will tell the story of Peter and the
Wolf. When he appears on the left wing of the stage, I wave
to him and call out loud: "Günter, Günter!"

He waves back and then begins to talk about Peter, a
small boy, who goes out of the garden into the meadow one
spring morning. The curtain opens and the musicians on the
stage make music of Peter, of the bird, the duck, the cat, and
Peter's grandpa. I look for the wolf, but he is nowhere to be
seen. The bird and the duck are quarreling. The cat wants
to catch the bird. Grandpa scolds Peter for having gone out
of the garden all by himself, into the meadow where the
wolf may be running around. The wolf can grab Peter and
eat him.

Günter tells the story, sometimes looking like the cat,
at other times like Grandpa. Suddenly the music becomes
ominous.

"Wolf" I cry as loud as I can. Yes, the wolf comes out of
the bushes, sneaking up to the little duck, catches it. I want
to run up to the stage to rescue the duck, but Mama holds
me tight.

Günter tells us that at the end Peter catches the wolf by
his tail and brings him to the zoo. I'm worried about the

duck. Günter says the duck comes out of the wolf's stomach alive because, in his haste, the wolf has swallowed it alive. I'm so happy that the little duck is swimming again in the pond and the wolf was not shot by the hunters. I like the story and I like the music. I think I can be Peter, strong and courageous. Papa says I'm a strong boy because I can knock over trees. Mama always says "Minochan, Minochan, *kashikoi ne!*"

NORTH SEA CURE

In the kindergarten many children are sick. Mama doesn't want me to go there. I already have a runny nose and a cough. Mama's thermometer shows I have high fever. Dr. Lezius says I may have the measles, which is spreading through the kindergartens. Mama says she had measles as a four-year-old. Papa says he had measles too. Mama feeds me chicken soup, and my fever goes down. She says if it's the measles, I'll get a rash all over my body.

The fever comes back and the rash appears. It's very itchy. I feel really miserable. Dr. Lezius looks at me and tells Mama I should take one-half of the pill, which will lower my fever, and in a few days the rash will start to disappear. I should keep warm, eat well, and sleep well; and I will be healthy again.

My rash has disappeared, but I'm now having pain in my ears, terrible pain. "Middle-ear infection," Dr. Lezius tells us, "happens often after the measles." He gives me antibiotics, but the infection may return, he says, may become chronic. A cure would be the North Sea. I should stay at the North Sea for four weeks breathing the ocean air. It's known, he says, that the ocean air does good. Tiny droplets of salt water are carried by the wind. The North Sea is rough. Its high waves sprinkle salty mist all over the coast.

Papa tells us he had horrible earaches as a boy. There were no antibiotics in the 1930s and 1940s. The only choice the doctors had was to pierce the ear-drums. In Papa's case twelve times. Lots of puss poured out. He remembers how painful it was. Nobody could bring him to the North Sea, because there was the big war. Papa thinks I should be breathing the North Sea air as soon as possible, before my condition becomes chronic. Dr. Lezius has pamphlets of children's homes along the North Sea coast and Papa starts calling different places. Most are booked, completely booked, but finally Papa makes a reservation for me. Mama stitches my name on all my clothes and underwear.

Papa takes three days off from the university. Mama makes a lot of sandwiches for our lunch on the way. We drive to Bremen, where the Bremen Town Musicians live. I want to meet the pony, the dog, the cat, and the rooster who make music, but Papa says we must stay on the Autobahn. We eat our sandwiches in the car and arrive at a harbor. From the pier, we go onto a ferry.

A ferry is a huge ship transporting cars and people. We are heading for Wangerooge, one of the smallest islands of the Frisian Island Chain. We come out of the ferry. The beach looks like a huge sandbox. I ask Mama if Mother Turtle will come to the beach. Perhaps not, because it's not a Japanese beach, where Urashima can meet her. Many white birds are flying in the sky. "Seagulls," Papa says.

We drive around and around, looking for the children's home where I'm supposed to stay. We find it, but the building is dark, looks unfriendly, and reminds me of a witch house where Hansel and Gretel are locked up in the Grimms' fairy tale. Papa and Mama hesitate to ring the bell. At this moment, the door opens, and an old woman appears to greet us. She has no bent back, no huge nose, and no red eyes; but I'm frightened. Papa takes me in his arms. Mama says "Wait here" in Japanese and follows the woman. After a while, she comes out and tells us this is not a good place.

We drive around and around, looking for some other children's homes listed in the pamphlets. Papa and Mama are not sure where to go, and it's already getting dark. We find a small inn to stay overnight. After supper, while Papa and Mama are discussing what to do next, I'm wondering whether that old woman we saw at the door was really a witch. Though she had no scarf around her head and was not riding a broom, she looked wicked. Listening to Papa and Mama I fall asleep.

Next morning Mama tells me we are going to see the pastor of the island. His name is supposed to be Zimmerli.

"Must be Swiss," Papa says.

After asking around Papa finds the church and the pastor's house. We knock at the door. His wife asks us to come in. They listen to our story and nod.

Mama asks: "Do you happen to know Emil Brunner?"

"Of course." The Pastor leans forward, suddenly very interested. "I was his student. How come you know his name?"

Mama tells them that Emil Brunner was a visiting professor at her university in Tokyo and that she took his class on Christian Ethics. "I was very close to the Brunners and often went to their house on campus. When Mino was six months old, we visited the Brunners in Zürich and stayed overnight."

The Pastor and his wife exchange quick glances. "We could keep Mino here with us for the summer. We have two sons, Kai, who is Mino's age, and Ilko, a bit older."

Papa carries my suitcase and bags from the car to a small room next to Kai and Ilko's. I meet the boys and like them. We run to the playground in the back of the house. Sand all over, very fine white sand. Playing with Kai and Ilko I forget about the time.

Finally, I go back into the house, because I'm hungry and thirsty. Papa and Mama are not there. They left without telling me. I cry and cry, looking for Mama and Papa every-where. Mrs. Zimmerli assures me they will come back to pick me up.

Why did Mama and Papa leave without telling me first? I cry and cry before falling asleep. Next day I play with Kai and Ilko. Pastor Zimmerli teaches me a long prayer about Dear Lord.

He says Dear Lord is in Heaven. He watches over me day and night.

I ask Pastor Zimmerli "When does Dear Lord sleep?"

He thinks about it for a little bit and says: "Dear Lord never sleeps."

Days go by, weeks go by.

Suddenly Papa and Mama are back. I'm so startled that I don't know, if I should cry or laugh. I put my arms around Mama's neck, never to let her go again. From now on I watch where they go and run after them everywhere.

Driving home to Göttingen, the feeling of abandonment overwhelms me again. I can't believe that Mama and Papa really picked me up and that we are driving home together.

Back in Göttingen, I run from room to room to convince myself that everything is still as I left it. All my friends are sitting in my bed waiting for me, all the books are in the bookshelf waiting for me, all crayons are there waiting for me, my Japanese drum and my magnifying glass too.

The tatami room is there with the long curtain, behind which I can hide; the large mirror, in front of which I dance with Mama; Grandpa's calligraphy roll on the wall; and Mama's desk.

I take a bath with my duck and frog, diving under water again and again, splashing till there is almost no water in the bathtub. After I'm all dry again, Mama cuts my hair short, clips my fingernails and toenails. She and I hang laundry together in the balcony behind the kitchen and I eat Mama's food which tastes so different from the food at the Zimmerli's at the North Sea.

Mama and I sing again. We listen to *Peter and the Wolf, Swan Lake,* and the *Nutcracker.* We dance to the music on the tatami, fall down laughing.

We walk into town and visit Dr. Lezius in his office, to show him how healthy I have become. We do groceries and buy a new pair of shoes because my old ones have again become too small. We go to the pony and Mama puts coins in the slot. I ride and ride till the pony stops.

Papa flies to Moscow for a conference. I help Mama cook, do the dishes, do the laundry and hang it on clotheslines, and ride my tricycle to Frau Fichtner's store to get rolls for our breakfast.

RAINBOW

Mama, Mama, how big is a whale? As big as Papa? Much bigger? As big as our tatami room? Still bigger? Eyes are small. Why? What do they eat? Only fish? What kind of fish? Do fish see the whales? Do they swim away from whales? Do whales eat Mother Turtle? No! She swims away and hides in her palace. How big is the mouth? Dark inside? Do they have teeth? Do whales have to brush their teeth too? No? Then lots of Karius and Baktus and holes in their teeth. Where do they sleep? Mama, but if they sleep in the water, how can they breathe? How big is a baby whale? Oh, so big! Does a baby whale drink mama's milk? But then water gets into the milk. Do mama whale and baby whale play hide and seek? Where do they hide in the water? Do they sing together? Oh, I want to hear their song.

Papa drives with us to Hannover, where he has to give a lecture. Mama wants to visit a museum with me. It's cold and raining outside. Papa drops us at the museum entrance. We walk from one large room to the next and look at the paintings hanging on the walls. Since I draw pictures with crayons, I want to see what some grown-ups have done. Mama stands with thoughtful face in front of some paintings.

"What's this?" I ask her, but she doesn't know what the lines and circles are. Some pictures are just a black square or a red dot on white. I'm bored and start to run around, calling to a guard, "Catch me, catch me!" Mama takes me to some other room where there are pictures of horses, moos, pigs, boats, birds, meadows, farm houses, and castles.

For the first time in my life I see a devil. He has horns and a *shippo*. He has a mean face. He looks wicked but different from a witch.

In the next room there is a naked man all in stone but he misses an arm. Another naked man had no penis. "Mama! He can't pee! We must call Dr. Lezius!" Mama says these statues were made a long, long time ago.

We go to the aquarium. This is a fun place. I like the slippery-looking eel. Mama reads on the board that this eel is an electric eel and that it uses electricity to catch fish. I wonder how an eel can produce electricity. Mama doesn't know.

All the windows of the fish tank are too high up for me, and Mama can't keep holding me for so long, because now I weigh forty pounds. I pull a chair from the corner, where a guard is supposed to sit, to the windows and stand on it. I watch a turtle, an octopus, a squid, and many fish.

"Mama, look! What do they eat? Where do they sleep?" Mama doesn't know, and the notice board doesn't give the answer either. In the last room we see the mammoth. Huge, with long, curbed tusks, but they don't scare me a bit. The mammoth is in the same family as the elephant, my baby elephant who sleeps with me.

Back in Göttingen, one afternoon, Papa takes me to his institute. I have been there many times and know his office well on the second floor. We climb up the stairs, holding hands and watching steps. I say hello to his students who are grown-up men, allowed to work in the lab. I'm still too small.

In his office Papa shows me a prism, which is made of glass and can turn sunlight into many colors. I can hold the prism in my hand. Papa shows me how to use it to catch a sunbeam coming through the window. There are old trees outside and the sun shines through their leaves. When a sunbeam passes through the prism and I hold a piece of paper behind it, the light changes into colors: red, orange, yellow, green, and blue. If I turn the prism upside down, the colors change and go from blue, green, yellow, orange, and red. Papa says the color sequence is always the same.

I look at the colors again and again, turning the prism upside down. I'm going to draw the colors on my paper at home and show Mama how the sunbeam looks like under the prism. She will be stunned.

Walking through the streets of Göttingen, Papa tells me that the colors of the rainbow are the same as the colors I have seen with my prism. "Next time when there is a rainbow in the sky, I'll show you," he says.

We are going to meet Mama at the corner. She is waving to us from a distance. I start to run. Papa follows me. We come to the fountain in front of the Concert Hall. I take a shortcut and hop onto the stone plates lining the pond. After a few steps, I fall into the water. Papa jumps in after me and picks me up. We are both soaking wet, I from head to toe and Papa from the waist down. Mama laughs. She walks home backwards, looking at us and laughing. Papa carries me in his arms, with water squashing from his shoes. Mama can hardly walk because of her fits of laughter.

At home, Papa and I take a hot bath, while Mama gathers our clothes and hangs them over the clotheslines. She cleans our shoes and wipes the wet traces we have left on the floor. Out of exhaustion she starts to be furious. She doesn't laugh

anymore but makes a somber face. She seems to have laughed
too much.

After dinner, I tell Mama that I can catch the rainbow
colors. She is taken by surprise. "Oh, Minochan, really? How
can you do it?"

I tell her, "Next time when there is a rainbow in the sky,
I'll show you."

While Mama sings at my bedside, I think of the colors,
red, orange, yellow, green and blue.

SWISS CHALET

Early one morning, we leave for Zürich. Papa can take a
ten-day vacation from the university. Our destination is Scarl,
a village close to Davos in the mountains of Switzerland. He
has arranged with his family friend, Dr. Heimann, that we
can stay in his chalet in Scarl.

From Zürich we drive to another country, Lichtenstein.
From Lichtenstein to another country, Austria, and from
Austria to Switzerland again. Mama teases Papa, saying he
doesn't know where we are going. After midnight we arrive
at Dr. Heimann's chalet. We go to bed immediately. I hear
tick-tock, tick-tock in the darkness. Papa says it's a pendulum
clock. I'm too sleepy to look at it.

In the morning, I look around. I see the tick-tock pendulum clock. Papa shows me how it makes tick-tock sound. I
see heads of deer with big antlers on the wall. They are hunt-
ing trophies—trophies everywhere, in all the rooms looking
down with sad eyes.

Mama doesn't want to look at them. She can't understand how someone enjoys shooting animals and decorates
the house with their stuffed heads. Papa agrees, but we can't
take the trophies down, because this is Dr. Heimann's chalet.

The sun is bright and strong, even though it's already
late September. In a small store in the village, we buy food,

including rolls, butter, and all sorts of ham for breakfast. We
like Swiss rolls and butter so much that we eat them all, and
Papa goes back to the store to buy more for lunch. We study
the map of the surrounding hills and mountains; it shows
where the trails begin and tells us how many hours we are
going to walk. Papa says we should start slowly and get used
to the altitude. I can hardly wait to see the meadow and the
hills with flowers, trees, insects, and birds, which I have seen
in my picture books.

We walk through the meadow along a trail. I see a little
yellow green thing jumping in the grass. I try to catch it, but
it jumps quickly away.

"*Batta-chan!*" says Mama. "Grasshopper," says Papa.
He catches one and shows it to me. *Batta-chan* has big eyes,
long, long legs, short arms, and a tail. No ears, one mouth.
Mama says there used to be swarms of *batta-chan* eating all
the rice in Japan. Papa says, during winter they survive in
the ground, then come out in spring and grow throughout
the summer. Carefully I hold one *batta-chan* in my hand. It
is frightened and keeps very quiet. I let it jump again and say,
"*Batta-chan, daijobu yo!*"

We walk and walk, then sit down to drink from bottles
that Papa carries in his backpack. We eat and continue the
trail. In one spot Papa finds mushrooms. "Porcino," he says.
"We can eat them." He collects the porcino in our empty
lunch bag. I join and bring all the mushrooms I can find to
Papa. He knows which ones are good and which ones poisonous. Mama says, he is an "expert" on mushrooms and one
time the Göttingen newspaper called him to identify good
or bad mushrooms people were bringing in. Mama is only
concerned with how to make spaghetti with porcino.

We walk still further up and find different mushrooms
under the trees. Papa points out those that are edible too.
Yellowish brown, called slippery Jack. Milk caps are red and

Mama's favorite. There are so many that we are running out of room in our lunch bag. When Mama again talks of spaghetti, we decide to return to the chalet.

Walking down the trail we are amazed how high we have climbed up. The air is getting chilly and we think of spaghetti with porcino sauce. Papa says we can also slice the mushrooms and dry them in the sun.

In the chalet Papa builds a fire in the fireplace. He stirs the fire with a long iron stick and pumps air with bellows. I throw pine-cones into the fire, one after the other. I watch them crack and burn.

During the next days we take different trails, find more slippery Jack and porcino, carrying them back, slice them nicely, and spread them in the sun. Sunny days and cool evenings. One day we hike up to the pass, 2300-meters high—fourteen hours up and down. Mama says, a record at my age of three years, eight months. On the way down, for the last hour, Papa has to carry me in his arms, because I fell asleep walking.

OPA AND OMA HOOSS

Opa and Oma Hooss in Marburg are going to celebrate their golden wedding anniversary. Mama and I lived with them and the whole family, Hans and Käthe, during our first winter in Germany, bitter cold. Göttingen was overrun by refugees from East Germany when the Iron Curtain came down and the Communists built a wall along the border. Papa couldn't find even a single room for us to live in. That's when the Hooss family in Marburg took us in. Mama packed presents for all of them. She also packed my inflatable blue pony. I'm going to pump it up and ride on it at the golden celebration.

There is no Autobahn from Göttingen to Marburg. Papa drives over country roads through fields and meadows on both sides. I'm interested in the moos. They are eating, walking, or lying down in the meadows. Last time in Marburg, I visited Uncle Otto who lives in a farmhouse with a moo, five pigs, one horse, and many chickens. He was plowing the field and I could walk beside the horse pulling the plow. "The horse is getting old," Uncle Otto told me. As the soil was turned, birds flew down to catch earthworms. I wasn't sure whether I should be happy for the birds or sad for the earthworms. In the farmhouse, I watched Uncle Otto cleaning the pigs' house and milking the moo. He promised me that he'll let me milk the moo when I turn six.

We arrive in Marburg just in time to get ready. I help Papa pumping up my pony and we all go to the community house for the golden anniversary celebration.

Opa and Oma sit at a long table on the stage. All others find their names on the tables facing the stage. Uncles, aunts, cousins, and other relatives of Opa and Oma are shaking hands with each other.

Uncle Otto from the farmhouse is among them. He lifts me up, saying "Already four and a half." I remind him of his promise: "When I'm six, you will let me milk your moo." He says, of course. He no longer has the horse, but now drives a tractor to plow the field. "What happened to the horse?"

"Ahh, retired." Uncle Otto taps my shoulder and hurries away.

Opa and Oma look lost on the stage, sitting at their table. Oma wears a flower garland with rose buds knitted into a golden lace.

"Opa and Oma have been married for fifty years," Mama tells me.

"Why do they look so sad?"

"They have stage-fright," Mama answers.

"What is stage-fright?"

"They feel uncomfortable sitting on that stage. Everyone can watch them."

"Opa! Oma!" I shout and wave to them.

Both smile and wave back to me.

One after the other the grown-ups make speeches, drink a toast in honor of Opa and Oma and finally we can start eating. Sitting between Mama and Käthe, I feel happy and eat everything on my plate. Last time I was in Marburg, Käthe showed me how to use a fork instead of a spoon. At home we always eat with chopsticks and never use knives and forks at the table. I find it easy, both ways, chopsticks or fork and knife. Doesn't matter, if the food is good.

After dessert, the entertainment begins. As the youngest in the hall, I'm asked to come out in front of Opa and Oma and sing songs. I start with *"Yuyake koyake"* and sing two more in Japanese and one in German. I get stuck with the second verse in German, but everybody helps by singing along. I'm happy because Opa and Oma are smiling and clapping their hands.

I ride my blue pony, large with round eyes, long ears, and a long neck. I ride and ride until Papa tells me it's time to go home.

TORRENT OF WHY'S

It's raining hard. Lightning and thunder.

"Does the Thunder god also come to Japan?"

"Yes. In the summer," Mama says.

I like the Thunder god. In my picture book, he runs through the clouds, stamping and beating three drums at once. I ask Mama: "The Thunder god and Dear Lord both live in the sky. Do they know each other?"

"Surely."

"Who is Dear Lord, Mama?"

"Dear Lord was once called Yahweh. He is the God of the Jewish people, but also of the Christians."

"Pastor Zimmerli says Dear Lord is in heaven. He watches

everybody on Earth. When people do something bad but repent, Dear Lord forgives them. When people do something good, Dear Lord writes it down. When people are sick and suffer, Dear Lord relieves them of their pain."

Mama says, "Pastor Zimmerli believes so."

"He says, if we die, we go to heaven to Dear Lord."

"Yes, he believes so."

"Where does Dear Lord go when he dies?"

"He doesn't die."

"Why not? Why not, Mama?"

"Dear Lord doesn't get old. He doesn't get sick. He is always the same. He never changes. He never dies. He is forever."

"But we all get old and die. Mama, isn't it so?"

"Yes, when we are old and sick, then we die."

"Aha, Opa and Oma Hooss in Marburg are old but not sick. Dr. Lezius is old, but not sick. He can carry me. I see... but why do they all go to Dear Lord when they die?"

"Because this is what the Christians believe. But there are other gods in the world too. They also help, like the Sun god who gives the energy to grow."

"What does the Thunder god do?"

"He brings wind and rain."

"Who is the Thunder god, Mama?"

"Long, long ago in Greece and Rome, also here in Germany, there were many gods. Some protected the farmers, some brought victories on the battlefields, some helped people earn a living and build houses. Some gods protected the people who had to travel. Others cured diseases. But the Thunder god was the king."

"Why don't people go to these gods after they die?" I ask Mama.

"Because those gods did not take care of people after they died. For those who died there were still other gods, the gods of the Underworld."

"Why the Underworld and not Heaven?"

"I don't know," Mama says, "but the idea of Heaven came with Christianity. Christianity taught the people that Heaven is the place where they go after death...but only if they were good. If they were bad, they go to Hell."

"Hmm ...Dear Lord is in Heaven. Does he have a big house—very big like the Concert Hall?"

"Much bigger. The entire Heaven is his house, so huge you can't see the end."

"But if so many people go to Dear Lord, it's soon going to be very crowded up there. Can't run around, can't play hide-and-seek, can't drive my tricyle. No fun to go to Heaven, Mama."

"Minochan, Minochan, we are not going to Heaven. We are here. We are together and do a lot of things together every day. When you grow up, you will become a Papa. Mama will become grandma and Papa will become grandpa."

"Oh yes, I will have many children, like in the kindergarten."

"Papa and Mama will come to visit you."

"And then my children are going to grow up and also become papas and mamas...then I'm going to be grandpa? Is that so, Mama?"

"Yes, it will be so."

"But then you will visit us, yes?"

"When you are grandpa, Mama and Papa will not be here anymore."

"Why not?"

"We can't live forever."

"Why not? Why not?"

"When we are very old, we'll die."

"No! No! No! I don't want to grow up. I don't want to become papa. I don't want to become grandpa! I want to stay always with Mama! Stay here, just here. Papa, Mama, Mino, we three together—always together. That's best. I want to be with Mama. I want to be with Papa. All three together, always."

VIGNETTES

Look, Mama, I can carry Papa's camera. It's very heavy.
  Don't worry, I don't let it fall.
  I'm taking pictures of you. Look at me, Mama, and smile.
  Don't worry, I'll not drop it.
  When I'm grown up, I'll buy a camera.
  So, look at me, Mama, and smile.

Quick, Mama, quick. Bambi jumped into the bathtub.
  All wet.
  It will catch a cold.
  Quick, Mama, quick, we have to dry Bambi in the sun.
  Oh, Mama, Bambi is crying, crying so loud.

Mama and Papa are going to the theater?
  I want to go, too!
  Why not?
  People just talk and talk on stage?
  Talk a lot of rubbish?
  Nothing exciting?
  Why do you go then?

I'm tired, my legs don't move.
  my belly hurts, I can't walk.
  I'm hungry. Mama, *onbu*!
  Carry me on your back, Mama, *onbu*!
  Ah, now I feel better!
  I'm not tired anymore.
  I'm so-o-o happy, Mama!
  Run faster, Mama, run.

Some day I'll go away, far away.
　　Mama won't cry, because I'll be a grown-up man.
　　Someday I'll go away, far away.

Now I'm a squirrel, Mama.
　　I must have a bushy tail.
　　Bind your scarf around my pants.
　　I live in the trees,
　　come down and hop around,
　　I pick up acorns and eat them with both hands.
　　I dig a hole and hide nuts for dinner.
　　I climb up the tree
　　and jump from branch to branch.

Mama, come here!
　　My squirrel tail is in the way,
　　Can't pee,
　　Come here, Mama, quick.
　　Oh, what a day!

Annette's finch died this morning.
　　Pastor Zimmerli said, when we die, we all go to Heaven
　　　　to Dear Lord.
　　But Annette's finch is not in Heaven.
　　Don't you understand, Mama?
　　I tell you once more, Annette's finch is not in Heaven.
　　No, no, no.
　　Annette's finch lies there in the cage. I saw it. Just dead.

I have dinner with Mama.
　　Roasted chicken—it's so-o-o good.
　　Papa is still working.
　　We keep one chicken leg for Papa.
　　But it's not going to be hot, when he comes home.
　　Do you warm it up for him, Mama?

Papa, why doesn't Mama talk with you anymore?
  Is she angry with you?
  Then just go to her and take her tight in your arms.

Mama, the flowers on the balcony.
  The sun shines too bright.
  I've put sunscreen on them.
  Yes, on the leaves and on the flowers.

Papa, come here. Can't sleep.
  My blanket is rumpled up,
  I'm not covered, that's why I'm still awake.
  Cover me nice and warm.

Mama, come here. Can't sleep.
  Teddy snores...can't sleep.
  Cat and frog are singing...can't sleep.
  Dog and Bambi are playing hide-and-seek...can't sleep.
  Talk to them, Mama, they should stop it.

Mama, come here!
  Duck is paddling with its feet.
  It tickles my belly...can't sleep.
  Baby elephant doesn't stop laughing...can't sleep.
  You must sing *"Nemure, nemure"* once more, till all my
    friends fall asleep.
  Then I can sleep too.

AMETHYST

Today is my birthday. I am turning five. Presents have come
from Marburg, Wiesbaden, Japan, and America.

  I have been thinking of giving Mama a gift. I have saved
many coins in my piggy bank, *bufubu*. I have earned the coins
for picking up our breakfast rolls from Frau Fichtner's shop

and bringing them home with my tricycle. I think I'm rather wealthy by now. I'll ask Papa to open *bufubu*, my pig with a slot on his back.

Mama, Mama, I would like to tell you a secret. Come here, I must whisper into your ear. I want to buy a ring for you, a beautiful ring. Everybody gives me presents. You should have a present too. How much does a ring cost, Mama? Oh, so much...I think I have enough money. *Bufubu* is very heavy, many coins in it. I can hardly carry it. I'm going to ask Papa to open it tonight.

I go to town with Papa to a small jewelry shop near his institute. The owner shows me crystals of different colors side by side and asks me which one I like best. I point to a purple crystal because that's the color Mama likes. Whenever we are in town, she stands in front of women's store with dresses displayed in the windows. Purple is the color she likes most. She never buys anything, but she stands there and looks.

The shop owner tells me the stone is called amethyst. This one is a particularly deep-purple amethyst. "It will be lovely, if set in eighteen-karat gold."

He seems to wait for my reply. I look at Papa. He nods. So, I tell the store owner: "I'll take it."

The ring will be ready for pick-up in one week.

I give the store owner all coins from my *bufubu*. He counts and counts. He shakes my hand and says, "Thank you very much." I see him exchanging glances with Papa.

I look at Papa because I'm not quite sure what that means. Papa says, "Oh, the jeweler promises to do his best."

I look at the store owner and tell him, "Do your best."

"Mama, Mama, close your eyes.

"Now open! This ring is for you.

"Put it on... Do you like it, Mama? The stone is called amethyst.

"You must wear it always.

"When you die, I will wear it for the rest of my life.

"Mama, Mama, why do you cry? Are you not happy with the ring? Don't you like the color of the stone?

"Don't cry, Mama. Smile."

FLOWER AS BRITTLE AS GLASS

Whenever Papa takes me to his institute, I see the students in the lab. I see the white billowing clouds that spread over their laboratory benches. They are pouring liquid nitrogen, but I am never allowed to come close. Papa says, liquid nitrogen is dangerous, and I should stay at a safe distance. But today, because it's Saturday afternoon and no students are in the lab, Papa wants to show me how to handle liquid nitrogen safely.

"You are already five and a half, old enough to learn it," he says.

I agree.

In his office Papa pulls an old lab coat out of a closet, dirty white and full of brown and yellow blotches. He holds it up to me and then, with a big pair of scissors, he cuts it short to fit my height. He measures the length of my arms and also cuts the sleeves to the right length.

"Liquid nitrogen is very, very cold," Papa says. "If it flows onto your skin, it boils off with lots of white smoke. Harmless. But if it stays in the hollow of your hand, it freezes your skin and hurts. It's dangerous. If it gets into your shoes, it freezes your feet down to the bone and destroys them."

He reaches into his briefcase and pulls out my sandals, which he has brought from home. "Take off your shoes and put these on. If the liquid nitrogen splashes over your feet and you wear shoes, it can enter and accumulate. With sandals, this can't happen. You're safe."

Next Papa gives me goggles that cover half my face. He fastens them tight around my head. "We must protect your eyes," he says. "The smallest drop of liquid nitrogen can blind your eyes."

Finally, he reaches again into his briefcase and pulls out two pairs of bulky leather gloves, one big and worn out for him. The other one is small and brand new. They are for me. We both put on the gloves and look at each other confidently.

"If liquid nitrogen splashes into your gloves," Papa says, "you just flip your hands downward so that both gloves flow off." He shows me how. I repeat it and it works. It's even fun. I try it again...*chaboom.*

Next Papa picks up a dewar and pours a bit of liquid nitrogen onto the floor. The nitrogen splashes and spreads out over half his lab, forming a dense, white cloud of very, very cold air. Then he gives me a smaller dewar to hold between my hands protected by the bulky gloves. Papa pours liquid nitrogen into my dewar, while I hold it. I see and feel the cold air flowing down as a white cloudy curtain. Papa tells me the moisture in the air turns into tiny ice crystals like the clouds high up in the sky, where it's very cold. He asks me to pour the liquid nitrogen from my dewar into a second empty one for practice. Then pour it back into the first. Liquid nitrogen has no color, no smell, looks like water.

It's really fun.

Papa takes some dandelions we have plucked on our way to the institute. I am very curious what he wants to do with them.

"Take one and dip it into the liquid nitrogen," he says. "Now pull it out."

The color of the flower has not changed. It is still yellow, but it's covered with white frost. Papa hands me a hammer.

"Smash it gently."

The flower fractures like a piece of glass. A thousand pieces, all over the table and on the floor, quickly softening again as they warm up to room temperature.

Papa lets me try over and over again with more dandelion flowers. At the end he says, "Now it's enough," and peels me out of my lab coat. He puts away the goggles and the bulky leather gloves and gives me back my shoes.

What I have learned this afternoon is that properties change with temperature. Even a soft dandelion flower can become as hard and brittle as glass, when I dip it in liquid nitrogen.

That is quite something. I must tell Mama. She will be stunned.

LITTLE NEWTON

We all drive to Cambridge, England. We are going to visit Dr. Josef Needham, an eminent biochemist, Papa says, and Master of the Caius College at Cambridge University. He is a great scholar and the author of many volumes of *Science and Civilization in China*. Papa has shown me the volumes in his bookshelf at home. Dr. Needham, now almost seventy, has been a member of the Royal Academy of Sciences for nearly forty years. He is fluent in Mandarin—reading, writing, and speaking. Grandpa knows him, too, and has corresponded with him about classical Chinese philosophy and natural sciences.

Dr. Needham has invited us to stay with him in the Caius College so that he can discuss things with Mama and Papa all night through, if necessary.

We are the only guests in the College, and we stay on the top floor. Very quiet, except for the chimes from the nearby church every quarter-hour. We sleep deeply through the first night and get an unusual breakfast—smoked eel with strawberry jam and toast. Mama is amused. Waiters with a white napkin over one arm serve us. They say "Sir" or "Madam" at the end of every sentence. We eat toast with butter and leave the smoked eel with strawberry jam on the table.

Dr. Needham welcomes us in his office. He is tall and slender, taller than Papa. He asks me my name and my age. Before he starts to talk with Papa, he asks Mama whether she would like to go out. We may take a walk through town, he says, for sightseeing and come back around noon. Mama thanks him but we all stay in his office.

The sun is shining through the window. The edges of the glass are cut at an angle. The beam of light passing through has colored edges. I take a piece of white paper from the desk, walk toward the window and catch the light. As I get closer, the colorful edges become thinner and thinner and disappear.

Dr. Needham must have watched me while talking with Papa and Mama. He asks whether I'm interested in light. I tell him about my experiment with a prism in Papa's lab and that I had seen the rainbow colors.

"Is that so?" Dr. Needham says. He fully turns to me, "I'll take you to the Trinity College Chapel, where we have a statue of Sir Isaac Newton." He goes on talking about Newton, who was here three hundred years ago, in this very same room. He observed light just as I was doing it, with prisms, and contributed to the modern theory of light.

Dr. Needham calls out to one of his students. "Here is a little Newton," he says. "Go to the Physics Department and borrow two prisms. Mino wants to study light."

Half an hour later the student brings back two prisms. They are larger than my hand and heavy. Dr. Needham takes me to a smaller room adjacent to his office. It has only one window, a narrow one, through which the sun shines brightly. He picks up a piece of cardboard leaning against the wall, large enough to cover the window frame. He punches a hole into it. Then darkens the room by putting the cardboard against the glass pane. Only a sunbeam enters through the hole. "Mino, here is your beam of light. Now play with your prisms." He leaves the door ajar.

I look around in the darkened room. Not much there except for one table and a chair. I push the table so that the sunbeam falls onto it. I pull up the chair and catch the beam with the first prism. I see the rainbow colors. Then I take second prism, placing it behind the first. Depending on how I hold the prism, the colors get either mixed up and whitish or more intense.

Dr. Needham comes back after a while, with Mama and Papa peeking from behind. I show him what I've found. He nods. He takes me across the hallway to another room, much larger, with students sitting at desks. "William," Dr. Needham says to the student who had brought the prisms, "get a projector for our little friend and set it up with a collimated beam."

"What is a collimated beam?" I ask.

"William, tell him," Dr. Needham says.

"It's a narrow and straight beam that we can make with a few lenses."

"All right, set it up for Mino," says Dr. Needham.

William takes me to yet another room, even larger, with rows of cabinets with glass doors. I can see lots of instruments. Some look like ones I've seen in Papa's lab. William takes lenses sitting on long rails and pulls a projector out of a bottom shelf. He gives me some of the pieces to carry. Very heavy. We make several trips. He sets up the projector and the rail with lenses. He switches on the projector, which smells a bit dusty, but the light is very bright.

William adjusts the projector so that the light passes through the lenses. He moves the lenses, three of them, back and forth. Finally, he is satisfied. "Now we have a collimated beam." He asks me to hold the first prism into the beam. With the sheet of white paper, I catch rainbow colors.

"Take those," William says, and he gives me two small stands on which I can put my prisms. I move the stands around on the table so that the rainbow colors from the first prism fall into the second prism.

"You've got it!" William says and leaves me alone.

I move the prisms here and there and watches the colors come and go. When I put one prism upright and the second prism with its top-down, the light coming out of the second prism suddenly becomes white again.

I didn't hear them, but Dr. Needham has come back quietly with Papa and Mama. All three are watching me quietly from behind.

"That's exactly how Newton did it here in Cambridge long, long ago." Dr. Needham's voice pulls me out of my concentration. He pats me on the shoulder and takes my hand. "Now we go to see Newton's statue in the Trinity College Chapel."

OBEY?

Papa has made an appointment with the schoolmaster of the elementary school closest to our house. He wants to register me for the coming fall. By then I'll be five years and eight months, old enough to be enrolled, Papa says, though the rules require I should be six.

The schoolmaster opens the door to his office, which smells of cigarette smoke. He is a short, gray-haired man. He lets us sit at a table and asks me my name, my birthday, where I was born, and where we live. He asks me to count the numbers from one to ten.

"All right," the schoolmaster says, nodding approvingly, "but you are not yet six."

"But I can speak German, English, and Japanese."

"Why Japanese?"

"Mama is Japanese. I've learned the Japanese alphabet and many kanji. Should I write some for you? Sun, moon, stars, mountain, river, trees—"

"No," the schoolmaster interrupts me. "They are of no use in Germany. Here we speak German and write German, do you understand?"

"I know fairy tales in German. I know some by heart—*Aladdin und die Wunderlampe, Hänsel und Gretel, Till Eulenspiegel, Kaiser Ohne Kleider*—all in German, and *Die Bremer Stadtmusikanten*, too."

"We don't teach fairy tales in school. School is serious, more serious than that."

I tell the schoolmaster: "I know how to catch the colors of the rainbow with a prism and turn them into a spectrum. I can make a collimated beam. I can also pour liquid nitrogen."

"School is not the place to fool around. Can you read and write?"

Now it's Papa who interrupts. "Isn't the school going to teach reading and writing?"

The schoolmaster doesn't answer him but asks me, "Do you know how to obey?"

"Obey?" I repeat his word because I have never heard it before.

"But you know the word *obedience*, don't you?"

"No," I say and look at Papa for help.

"Are you ready to obey your teacher in school?"

I don't know what to say. Papa looks irritated.

The schoolmaster rises from behind his desk. For the first time, he smiles but it looks more like a grin. Turning to Papa he says, "Your son is not yet mature enough for school."

We are dismissed.

MY BECHSTEIN

Mama thinks I should learn to play some instrument, because, she says, I have an ear for music. From her friends at the Deutsches Theater she gets the name of a piano teacher, Miss Venus. She teaches in a garden house not far from Baurat

Gerber Street. I sit with Mama on a bench in the bright, sun-filled studio surrounded by a flower garden. In the middle of the room stands a grand piano. Miss Venus smiles and asks me my name. She asks me to sit next to her at the piano and touch any keys I like. I'm fascinated to hear the sound I can make. I want to continue and make more sounds. She always breathes deeply before starting a sentence. Though I am not yet six years old, Miss Venus is not much taller than me but she is a grown-up. She has twinkling blue eyes and curly blond hair. Miss Venus tells Mama that I can start any time. We decide that I'll go to her for lessons twice a week.

Papa has found a grand piano for me in a piano store in town. We all three go to look at it. It's chestnut brown. The store owner is giving it three new legs. It's a Bechstein, almost one hundred years old, built in 1870 by Mister Bechstein himself. It has his signature written on the inner frame, where the strings are. Papa has done some research. He says the Bechstein B Grand is one of the finest instruments built at that time.

Mama asks the owner about the history of this piano. "Where has it been for the last hundred years?"

The owner mentions it once stood in the Hohenzollern Castle.

"Hohenzollern Castle? When?" Papa interrupts him.

"Don't know really, but for many years."

At home, after dinner, Papa and Mama talk about Hohenzollern. It's one of the most majestic castles in the whole of Europe, the ancestral home of the Hohenzollern imperial dynasty, which dates back well into the Middle Ages. It sits on a mountaintop in southern Germany. Papa's mother grew up there because her father, a Prussian nobleman, was the castle's commander before World War I. Now it's clear we are going to buy this grand piano.

Mama brings her lexicon and shows me a picture. It's just

like in the fairy tales—a large gothic castle on the top of a steep mountain. I've built such a castle with wood blocks and with Legos, but I never knew Papa's mother had lived there as a child. Did she climb to the top of the towers and run along the fortification walls? Now I'm sad that she died when Papa was still young. I can't ask her. Papa tells me that his mother's father, the castle's commander, my great grandpa, was very much into books and music. He wrote poems and played piano—surely this Bechstein. He was an outspoken critic of World War I but had to serve as an officer at the front. He was killed in the war.

The piano is very expensive.

"Don't worry," Mama says, "my essays are now published every week in *Die Zeit*. We have money to buy the Bechstein."

I practice at home. Twice a week Miss Venus gives me lessons. Often she invites me to her half-timbered house in Friedland, twenty minutes by train from Göttingen. I stay there with her and her sister Mrs. Schöne, also a pianist. The evenings are getting cold and we put wood into the stove in their living room. I recite the story of Alibaba, the bandit, while the fire in the stove is sizzling. They listen to my story with *aahs* and *oohs*. At the end, they take me tight in their arms. We visit their brother, an organist. He shows me how he plays the organ. With hands and feet. He explains the reed pipes and the swell boxes. He plays a piece I like very much. He says it's Bach. Miss Venus likes Bartók. Mrs. Schöne is more for Beethoven. I like Beethoven too.

Christmas is near. Papa and I drive to the forest beyond my old kindergarten, and we cut a big pine branch. At home we decorate it with origami I'm folding with Mama—cranes, boats, flowers, cats, and samurai helmets. With my color crayons I draw on a paper Papa, Mama, and Mino walking together against a background of evening sky. One larger figure walks behind us. I don't know who this person is, but Mama keeps my drawing in her diary.

———

The concert is announced for May 4th, 1968 at the Music Hall of the College of Education in Göttingen. Fifteen boys and girls from the beginners' class gather backstage surrounding Miss Venus. We are from age six to eleven, all nervous, some hiding behind others, and most of us very quiet, listening to the instruction of our teacher. Miss Venus is smiling all the time, telling us we are going to do fine. She calls us one by one in the order of appearance and asks us to sit and wait on a long bench in the empty backstage room.

I am the third on the program. I come out to the stage and see many people in the hall. Papa and Mama are among them. They wave at me. I go to the piano in the center of the stage, climb onto the stool and realize that I have forgotten my music sheets somewhere backstage. I remember the melody of the children's songs I'm supposed to play. So, I just produce some sounds from memory. The first trial is not perfect. After a little while, I start anew and play as far as I remember, stop, think for a while and finish the rest.

I look at the audience. Silence. It was too short. Everybody is waiting for more. I make my own sounds, look at Mama and Papa, and make more sounds. Still not enough? I start with *do, re, mi, fa, sol, la, ti* and *do*, and come back down *ti, la, sol, fa, mi, re, do*. I repeat just like I do before lessons with Miss Venus. Then I make all sorts of sounds— first with one finger, stop, repeat; then different sounds with both hands on a different key, and stop. I stand up. Mama and Papa clap and smile. The others follow a little later.

When I come backstage, Miss Venus is wiping her tears and tells me, "Oh, Mino, what a music." I'm not sure what she means, but the second time I go on stage, program No. 7, everything goes exactly as practiced. I walk to the

stage with Marianne, a quiet, ten-year old girl. She carries
our music score and it is for four hands. It's a children's song
from Sweden and Marianne plays very well. I dangle my legs
because I can't reach the pedals yet. The audience applauds
immediately after we have finished.

All afternoon the concert has filled the hall with melodies of children's songs from Germany, Sweden, France, and
many pieces by Bach, Mozart, Haydn, Beethoven, Schubert,
Schumann, and Bartók.

At the end, we all appear on stage and surround Miss
Venus, who tells the parents in the hall how well their children have learned and performed. A long applause. The
oldest student comes on stage with a large bouquet of flowers and presents it to Miss Venus.

MIDSUMMER NIGHT'S DREAM

In the summer of 1968, we land at Miami Airport. The
moment the door of the airplane opens and we step out,
air hits us in the face—hot, humid air. Papa rents a car and
we drive to the Briny Breakers, a family-owned inn along
Pompano Beach. Mama got the name from her Penn State
friend, Mrs. Sieg, who comes here every winter. The highway
is empty, the inn is vacant, and the beach looks deserted. We
are the only guests. We get the best room with the best view
of the ocean.

According to Mrs. Sieg the place is full during the winter
season, when masses of New Yorkers and Northeasteners
exchange the cold for the Florida sun. In the summer, however, almost nobody is here because it's so hot and humid.
Only big expensive hotels have air conditioning.

Mama is beaming with joy. She loves the heat. It reminds
her of summer in Kyoto. Germany is far north, she says. Most
people forget how far north, the same as Newfoundland
and Alaska. The sun is weak and there is hardly ever a true
summer day.

At Briny Breakers the waves are making woooshing sounds right in front of our room. We forget how tired and hungry we are after the long flight. Our full charter plane from Düsseldorf to New York had to stop in Iceland and Newfoundland for refueling because it was an old propeller plane with tremendous whir and constant shaking. On the flight from New York to Miami we could recover from our exhaustion because the plane was a quiet jet, nearly empty.

Though we are going to stay here for the next five weeks, we shouldn't waste time sleeping in the middle of the day. We hop into our bathing suits and run out to the beach. The sun is burning hot, but the breeze is cooling. We run barefoot over the fine white sand, splash into the water, and feel the waves. We can't believe that there is such a beautiful place in this world. The beach is deserted. Only palm trees. No traffic sounds. Only the waves.

Mama starts to do yoga on the beach and Papa takes me by the hand. We walk deeper into water. This is the first time I really walk into the ocean as I did a hundred times in my imagination playing Urashima on the back of Mother Turtle. My feet touch the soft sand and the waves come higher and higher. The water is so-o-o warm. Looking back toward the beach, we can see Mama, but she has become tiny. Now the surf begins to reach my neck. We walk back to shore. Mama and Papa say we need to buy flippers and sunscreen.

In the following weeks, I see many marvelous animals and plants, and we do lots of things we've never done before. For instance, I go with Papa to a nearby harbor to buy fish. People here do fishing for sport. They catch fish, take photographs of themselves with the dead fish and then throw them back into the ocean. The bigger the fish, the prouder the fisher. Papa buys some fish from them. Back in our kitchenette in the Briny Breaker, Mama makes sashimi, but the fish has no flavor. Without wasabi and soy sauce, no sashimi tastes good, she says. She fries the fish. Still little flavor.

We stop buying fish at the harbor. Instead Papa and I go snorkeling. Mama is afraid of the water. I want to go with her underwater and see all those wondrous colors of fish with her, but she can't put her face in water. She can't snorkel. She stays in the shallow water until the waves touch her belly bottom, and then runs back to the beach. She once told me that when she was a little girl, an uncle of hers had thrown her into a fast flowing river, so that she would learn how to swim. Instead she almost drowned. Since then nothing has changed her mind—she is afraid of water. I'm lucky. I don't have such a bad uncle. I'm now able to swim. I have flippers; and with them, I can snorkel as much as I like.

Once I was in shallow waters near the harbor but without flippers. I just wanted to see some starfish. The tide was turning and the current suddenly carried me out to the open ocean faster than I could swim. Papa spotted me and ran along the pier to get ahead of me. Then he jumped from high up into the water like a torpedo and swam towards me, mostly underwater. After a few more strokes he caught up with me. I was really happy, clinging to him. He didn't try to swim back toward the harbor, against the current, but instead swam parallel to the beach so that we could slowly come out of the current. I kicked my legs too, and we swam together for quite a time, until we reached the beach, far, far from where I had entered the water.

Every night we take a long walk on the beach. The air is still very warm and the waves are gentle. One full-moon night we see a large turtle on the sand beach, some distance from the edge of the water. Mama says, *"Kame-chan!"* I say, "Mother Turtle!" Papa says. "A loggerhead." As we gently approach the turtle, we see that she is fiercely digging the sand with her hind flippers, preparing a hole in the sand. "It's to deposit her eggs," Papa whispers.

In the moonlight we see Mother Turtle shedding tears. Papa tells us she has to shed tears to keep her eyes from drying out. If they would become dry, she could become blind. I am happy she is not weeping. The full-moon light is bright and blue. Mother Turtle looks greenish blue. We sit on the sand praying for her and for her eggs. Now she is finished. She carefully shuffles sand over her clutch with both hind legs and then moves back toward the ocean.

I always thought Mother Turtle must be large to carry Urashima on her back to her coral palace at the bottom of the sea, but I never imagined how large she actually was. I could have lain down flat on her back, stretching my arms and legs, and still she would be larger than me.

I walk by her side, as she is almost swimming over the sand. I tell her that I know where she is heading, that she will swim deep into the ocean to her coral palace. I know it, because I have been Urashima.

A few days later, we discover hundreds of tiny turtles, hatchlings from the egg clutches some other sea turtles must have deposited in the sand. The hatchlings are running as fast as they can towards the water's edge. They look so small and helpless. Frigate birds are patrolling in the air above. They dive down and snatch the baby turtles from the sand. They pick up one after the other and swallow them in midair. I try to chase the frigate birds away, but they just ignore me. They continue to snatch baby turtles off the sand just in front of me and faster than I can turn around.

Papa and Mama join me picking up as many little turtles as we can carry in our hands and bring them to the water. We run back and forth many times, but even those baby turtles that we have put into the water are not safe. We see how they try to paddle as fast as they can away from the beach, but the frigate birds continue snatching them out of the water. No papa turtle or mama turtle is there to protect them. I am very angry.

One day we drive to Miami Ocean World. A big place with many animals. Ducks, ibis, geese, herons, and egrets, alligators of different sizes, sea otters, even a manatee, and loggerhead turtles. We watch the performance of the killer whales; but Papa says it is cruel to keep the orcas in a pool, because they are the fastest swimmers in the ocean, crossing every year thousands of miles from the Arctic to the equator.

I love the dolphins. They have cute smiling eyes and look cheerful. I want to swim with them in the ocean.

Mama loves the pink flamingos. Hundreds of them are standing in a shallow pond and they look like the evening sky after the sunset.

In the last week of our vacation, Papa takes me to the Everglades National Park. Mama has been galley-proofing her two books, which will appear in the coming fall in Munich. One of them is the collection of her *Zeit* columns, and the other is the complete German translation of the oldest Japanese literature from the tenth century. She already sent back the first galley proofs to Munich and is now working on the second and final revision. The two publishers have been waiting, calling her here in Florida. So, she can't come with us.

After a long drive, we arrive at a side entrance of the National Park. Only a few cars are parked there. A ranger lets us park there and we get a canoe with paddles. We put a lot of mosquito repellent all over our skin. It is humid and, hot. No breeze from the ocean.

I sit in front and Papa sits in the back. He says he has never used a paddle in his life, so he is going to find out how it works. I have seen in one of my picture books Indians paddling a long canoe. There are always two with a headband and a feather. Using both hands, one paddles on the right side of the canoe, the other on the left side. I tell Papa, we should paddle like the Indians. Somehow we make a progress in a canal, between a network of sawgrass, vines, and mangrove forests.

After a while we face three wide canals spreading in three

different directions, and again more canals and swamps. We paddle here and there and see fish, colorful birds, and one atrocious-looking big alligator. After a long paddling, we get tired and turn around to go back.

Papa looks worried, saying he is not sure which way we are supposed to go. Every corner looks alike, and we can hardly see where the sun is. I remember one corner with bright pink and white flowers. I remember we turned left there. I tell Papa we should find that flower corner and turn right. It takes us hours to come out of the labyrinth of canals. Night is falling. It's already pitch-black by the time we arrive at the parking lot. No ranger. No other car except ours. The time is past eleven o'clock at night.

I haven't seen any puma, any black bear, any opossum, any raccoon, or any bobcat in the Everglades, though I've read about them in the pamphlet. Mama was very worried. She says we must have gone to a wrong entrance or seen a midsummer night's dream.

YELLOW PERIL

Papa has been appointed professor at the University of Cologne. He is going back and forth between Göttingen and Cologne to check his new institute at the university and to find a place for us to live. Mama has been busy packing.

Ever since I started to walk, she has measured my height and drawn a line on the doorjamb. She did it whenever something memorable happened, such as when I cried because moo came into my dream, when I sang *"Sho sho shojoji"* all by myself, when I tried to see Karius and Baktus through my magnifying glass, when I dialed the police to report Mother Turtle swimming in our living room with Urashima on her back, when I got my tricycle and peeled a potato, when I saw "Peter and the Wolf," when I came back from the North Sea, when I made rainbow experiment in Papa's lab, when we came back from Dr. Needham in Cambridge, when I thought about Dear Lord and the Thunder god, when I bought a ring for Mama, when I had my debut at a piano concert, and after our Florida vacation.

Papa found out that all the pencil marks Mama had made on the doorjamb were just thin straight lines. Each with exact centimeters and millimeters of my height, each line labelled "Mino...Mino...Mino..." But with no dates.

"What did you do?" Papa said to Mama, clearly perplexed. "What did you do? No dates."

Mama was angry with herself, wandering from room to room, scolding herself, "*Baka, baka, baka,* no dates." I had never seen her so upset, so close to tears. I took her in my arms and told her, "Can't be changed, *Shoganai yo.*"

How good, Mama kept her diaries!

Before we move to Cologne, Mama and I once more walk in the old town of Göttingen, with its half-timbered buildings, mostly family-owned stores for generations, where we have shopped for ham and sausage, books, my shoes and clothes, toys, household goods, bedding, medicines, vegetables and fruits, film for Papa's camera and the amethyst ring. Mama looks at the shops, talking to herself, "'Goodbye' is such a sad word." We visit Dr. Lezius in his office. He is surprised to hear that we are moving to Cologne, but happy that, since my stay at the North Sea, I haven't gotten any earaches again. I give him a blue balloon as my farewell gift. We walk to the Concert Hall and see the fountain where I once fell into the water. We look at the Deutsches Theater, where Günter and the other actors play.

We walk up the Baurat Gerber Street and look up at our apartment with the green balcony. How often have I watched from this balcony the moon shining in the black night sky. We go around the corner to Frau Fichtner's store to say goodbye, then walk to Miss Venus's garden house. Marianne is just finishing her lesson. I wish her success for the next concert.

Miss Venus holds me tight and promises that she will come to visit us in Cologne.

Our house in Cologne is two blocks away from a well-tended city park and within walking distance of the streetcar terminal and bus station. The public school in our suburban neighborhood is reputed to be one of the best. Its name is "Albert Schweitzer School" after the famous doctor and Nobel Peace Prize winner.

I'm happy and curious to go to this school. Papa drops me in the morning at the school gate and I walk home after school. Our teacher is Mr. Schneider. He is very strict. He always wears a stern expression on his face when he asks us to do this and that. He always ends his sentence with "Understood, children?", "Get it done, children," or "Don't waste time, children."

At ten o'clock we get a fifteen-minute break. All classes are over by one o'clock in the afternoon, and the school closes.

I read the textbook that Mr. Schneider has given us. It is quite boring. Now I must learn how to spell so many words. Mama is still busy unpacking, cleaning, and cooking. She has to respond to invitations from different bookstores and TV stations. She also must answer many letters from readers.

After a few weeks, three girls from my school join me on my daily walk home. Erika, who is already ten, and her sisters, Monika and Angelika, who are twins. They live not far from our house. We often like to take a trail that runs through the park forest. The trail is lined on both sides by white birch trees. We run a little race and then slow down again to chat about games we may want to play at home. Monika and Angelika talk about afternoon TV programs for children. Erika has homework to do, but the twins are going to watch *Pinocchio*. I tell them we don't have TV at home. Therefore, I've never seen *Pinocchio*. Doesn't matter. I have so many other things to do in my new room. I've already unpacked my books, toys, and records.

All of a sudden, several boys jump out of the bushes between the birch trees. The biggest boy plants himself in front of me, blocking my way. There are five of them, as far as I can see with a blink of the eye. "Chinese or what?" He punches me into the face. Another one kicks me in the behind. I hear them shout "Dirty *chink*…Get out of here…Go home, dirty *chink*!" They surround me tightly. I can't escape. One of them spits into my face, yelling, "Black-hair devil…dirty *chink*." My backpack is ripped off my shoulders and my anorak torn. They kick me again and again, even after I've fallen. I hear Erika cry, "Stop it!…*Stop it!*"

At the end, while I lie on the ground, nose bleeding, Erika bends over me asking, "Mino, Mino, are you all right?"

I come home with my badly torn anorak and without my backpack. Mama is busy in the kitchen preparing lunch. I quickly go upstairs to wash the blood off my face and comb my hair, then come back down and sit at the table. Mama is still in the kitchen, telling me how well her two books are doing. She doesn't look at me and doesn't notice my face is swollen and all messed up.

I feel uneasy. I'm ashamed. These boys kicked me, beat me up, and spit on me—only because my eyes are like Mama's and my hair is black like hers.

Mama tells me from the kitchen that she's invited to appear on public TV for one hour. I'm happy for her but feel miserable about myself. I don't want to have lunch. As I stand up, I feel the pain all over my body. Limping, I go to my room upstairs. I lie down on my bed. I'm ashamed. I want to forget.

Mama comes and sees my face. "Minochan, Minochan, what has happened?" She kneels at my bedside, looks closer at my face and runs to the bathroom. She gets a soft towel and alcohol to wipe the spots where I'm bleeding. It hurts. "What has happened?"

I tell her about the boys who attacked me and the girls who tried to protect me but couldn't.

Mama helps me undress. She sees the blue spots, where the boys have kicked me. "Oh, no! You're blue all over your body." She touches the sore spots and I squeal. She gently rubs ointment where my skin is blue and swollen. "Do you know these boys?"

"No, never seen them before, but they are from the Albert Schweitzer School."

The doorbell rings. Mama runs down. It's Monika and Angelika. Mama is talking to them downstairs, asking and asking. The girls are answering, but I can't make out any words.

When I wake up, it's already almost dark outside. I must have fallen asleep. Mama sits on the chair next to my bed, dozing. I see my backpack. For a moment I think I must have dreamed, but, as soon as I move, I feel pain all over my body.

Mama opens her eyes and leans over. "Monika and Angelika brought your backpack. They told me there were five boys from Erika's class and her parallel class, all ten years old. She knows the names of two of them. I've written them down." She takes me in her arms and gently puts her face against mine. "Monika and Angelika told me that the boys called you a *chink*... I asked them what it means, but they wouldn't tell me. Something to do with Chinese. Too bad Papa isn't here. He's coming back only in three days."

In the morning Mama insists on coming with me to the Albert Schweitzer School. She asks to see Mr. Damm, the principal. We have to wait for an hour. Finally, Principal Damm has time to receive us. Mama tells him what has happened. She points at my face and shows him the blue spots on my body. She mentions the three girls Erika, Monika, and

Angelika as witnesses. She gives Mr. Damm the names of two of the five boys who attacked me.

When Mama has finished, Mr. Damm leans back in his chair and asks her what she expects him to do.

"These boys are from your school. They attacked Mino. They called him a dirty *chink*. You have to do something about it."

"I can't do anything unless you bring these boys to me."

"How can I bring them to you? You are the principal. You know where they are. I gave you two names. You can surely find out the other names."

"Actually, I'm a wrong person to handle this case," Mr. Damm says without moving in his chair. "Whatever happens after school and outside the school premises does not fall under my jurisdiction."

I see Mama getting very angry. She leans forward and looks straight into Mr. Damm's face. "They called Mino a dirty *chink*. Do you know what dirty *chink* means? Are you not upset?"

"Many things happen off the school premises. Traffic accidents, bad falls, broken bones, and scuffles among youngsters. Things happen all the time but when they happen outside the school, I can't be held responsible."

"But *chink* is a racial slur. Boys from your school are calling Mino dirty *chink*. You, as the Principal, should step in."

"Well, your son...well...his hair is black and his eyes... how should I put it...are different from the eyes of our children. Your son looks different, doesn't he?"

"That's not a reason for beating him up."

"Boys often have too much energy and must get rid of it somehow. The boys probably said *chink*, because they don't know the difference between Chinese and Japanese. You're from Japan. Even for us adults it's difficult to tell Chinese and Japanese apart."

"That's not the point," Mama replies sharply, "why 'dirty'? Why 'dirty *chink*'?"

"That's the personal opinion of these boys. I must respect it."

"If they had said 'dirty Jew' and beaten a Jewish child, would you also stand by and do nothing?"

I see the principal frown. He sucks in his lips so that his mouth turns into a line.

"I know, times have changed in Germany," Mama says. "Germans no longer spit on Jews, beat the Jews, kick them and murder them."

The principal abruptly stands up, a stout man with thinning hair and a square face.

Mama continues, unimpressed: "Now you're looking for another target for your racism," she says.

The schoolmaster walks to the door and opens it. "I have been made aware you're writing in *Die Zeit*. Very liberal, clearly leftist," he says. "I've read some of your stuff. You're criticizing Germany." He gestures us out with one hand and shuts the door behind us with a bang.

Mama and I leave the school. Down a block, on the other side of the street, I see Mr. Schneider, my teacher. He comes over to talk to Mama. He already knows that I have been beaten up and why I couldn't come to class for a day. "Hello, Mino, are you all right now?"

Mama tells him that we have just come from talking to Principal Damm and that it was a very disappointing meeting. She tells him why.

"Not surprised," Mr. Schneider says. "You may not know it, but Mr. Damm is one of those Nazi hang-overs, a former card-carrying party member. He was a small enough fish to have gotten away with it after the war. Anyhow, next spring he's retiring."

Until Papa comes back, Mama takes me every morning

to school and picks me up at one P.M. After that Papa drives
me to school in the morning and Mama walks home with me
after school. We don't take the trail through the forest but
stay on the sidewalk along the street.

Mama and Papa still talk about what happened to me,
but I want to forget about it. I don't want to be scared all the
time that the same could happen to me again. One afternoon
Papa and Mama come back from the post office, which is on
the other side of the forest and the park. Mama's face is all
dirty and full of dust.

I've never seen Papa so upset. He walks in circles in the
room stamping the floor. "That's impossible...always the
same...it never ends...."

Slowly I get it. Papa and Mama were walking back
through the open park, on a wide gravel trail, when several
boys came from the opposite side, about twelve or thirteen
years old. Papa and Mama were talking to each other and
didn't pay attention to these boys, but suddenly a load full
of sand and gravel hit Mama's face.

The boys were mocking and jeering: *"Chink...chank...
chunk...chink...chank...chunk...dirty chink!"* Papa tried to
grab one of them, but they all ran away faster than he could
catch them, laughing and jeering.

"I was walking right next to Mama," Papa says, "and
nevertheless this happened. That's unacceptable. What country is this?"

"It's your country," Mama says, with bandages on her
face.

"This shows what the parents of these boys are talking
about at home. That's where these boys get the ideas."

"But why Chinese and Japanese?" Mama asks. "There
are almost no Chinese and Japanese in this country of sixty
million Germans. Maybe six hundred Japanese in all and even
fewer Chinese. Most Germans have never seen a Japanese or

a Chinese in their lifetime. Why do they call us *dirty chink, chink, chank, chunk?* We haven't done anything to them, surely nothing bad. Japan and Germany were even allies during the war."

"Well," Papa says, "Attila the Hun invaded Europe in the fifth century, and Genghis Khan's army swept through in the thirteenth century."

"That's far-fetched," Mama says. "So many centuries ago. No one can keep hatred for that long."

"Yes, they can. People remember," Papa replies. "If they have been victims, if they have suffered, they don't forget. Even after centuries." He is getting all revved up. "When Japan defeated Russia in 1905, the German Emperor coined the term *Yellow Peril*. It became a famous quote, stirring up old and deep emotions. Though Europeans have been waging wars against each other for centuries, they suddenly felt united against the Yellow Peril."

I am hungry. Mama jumps up and runs to the kitchen. While eating dinner, Papa seems to be lost in thoughts. At the end he says, "Forget the Albert Schweitzer School. We are going to find a private school for Mino—the very best."

CRANS-SUR-SIERRE

After driving for two days we arrive in the French-speaking part of Switzerland. The road up to Crans-sur-Sierre is long and winding. Papa finds the inn, where he has reserved a room for us. It's near the boarding school where I'm going to enroll, if we like it.

It's already sundown. We are hungry and tired. Through a large window in the dining room, we can look out over the valley in front of us and see the high mountains on the other side. On the left I see a white four-story building. This looks like the school, École Chaperon Rouge, in the photo in the brochure.

I sleep between Mama and Papa. In this way, I feel protected. The school has been recommended by the Rector of the Collège Calvin in Geneva, where Papa got his baccalaureate. Like in Papa's Collège Calvin, the language in the Chaperon Rouge will be French. After Japanese, English, and German that's fine with me. Just one more language.

After breakfast, we walk over to the school. The air is cool. Green everywhere. Reminds me of our vacation in Scarl at the Heimann house. Perhaps we can find a lot of mushrooms here, too. Papa talks about his college days in Geneva. Mama is quiet, holding my hand all the way.

At the entrance of the Chaperon Rouge, Monsieur Bagnoud greets us with a big smile and a melodious stream of words. He is the director. Papa answers in French. Mama says in English that she is glad to see him.

Bagnoud looks at me and asks whether I want to see my room. Papa translates into German what the director just said in French, and I reply, "Oui, Monsieur." Bagnoud explodes into a new stream of words that I don't understand. He opens his arms and takes me in. He walks with me down the corridor, gesturing Mama and Papa to follow us.

My room is on the second floor with a spectacular view over the valley and the mountains beyond. Monsieur Bagnoud tells me that I'm going to share the room with Pascal from France, also six years old. There are seventy children altogether. They come from thirty-eight different countries. Many are children of diplomats at the United Nations in Geneva.

We meet Madame Bagnoud, who is a nurse. If I don't feel well, she says, I should tell her. If I get ill, she will take me to the hospital down in the valley.

Monsieur Bagnoud introduces me to a young Swiss woman, Monique, who will take care of me and Pascal. She will make sure we wash our faces every morning and evening, and that we take a bath twice a week. I should call her

Monique or Mademoiselle. She speaks only French. So I must learn French quickly.

Before leaving home, Mama stitched my name on all my laundry and embroidered MINO on the back of my new jacket. Monique puts them one by one into the closet which is going to be mine.

Lunchtime is noisy. I have never seen so many children eating together in one room. There are several mademoiselles, who scold the children who are not behaving well. Everybody seems to talk at the same time. It's hard to keep French, German and English apart. The food is delicious and plentiful. A small chunk of spinach comes flying through the air and sticks to my face. It's from a boy at the next table. He waves and calls out. *"Salut, Mino, c'est moi, Pascal."* That's how I meet my roommate.

Papa and Mama will stay in the inn for the whole week to see how I feel.

In the afternoon I'm already assigned to the German class under Miss Hischier. She goes from desk to desk checking each assignment. She is gentle but firm. There are three divisions in her class. I'm in Division One. I look through my textbook. Lots of stories which I am anxious to read. Miss Hischier is surprised to find out that I can already read books but can't yet write all the words. "You'll learn spelling quickly," she says, patting my shoulder. "You'll have to write letters to your parents every Saturday."

I look at Papa and Mama, who sit on chairs in the back of the classroom. I feel strange that I should write letters to them.

Mama chats with Miss Hischier and finds out she is from Bern in the German-speaking part of Switzerland. She wants to visit Japan someday. Her father's brother lives with his family in Tokyo, where his Swiss company has opened a branch. Miss Hischier asks Mama which books she would recommend about Japan. Mama promises to send her a few books.

There are twenty-three boys and girls in Miss Hischier's German class. My next class is arithmetic. Then comes French.

Wednesday afternoon is free. We can play inside or outside. Monique and the other mademoiselles join us playing. But today I want to go out with Mama and Papa. Taking one of the many trails, we walk uphill through dense pine groves. Papa takes pictures of me with Mama and she catches pictures of me with Papa. We run a race uphill until we have to sit down on the grass. There are cows with big bells around their necks. We find slippery jacks and other mushrooms, which we gather. Mama sees her favorite milk caps. We can't possibly collect them all.

Mama brings the mushrooms to Mrs. Dick, the inn keeper, and asks whether she could sauté them for dinner. When we come back, Mrs. Dick serves us spaghetti with sautéed mushrooms with heavy cream and black pepper.

I am becoming familiar with the Chaperon and I like it. Monsieur Bagnoud assures Mama, Papa, and me that I may come to see him anytime with questions or problems. Mama thinks he must be in his late thirties. "He looks like a genuine teacher," she says, "who knows what he is doing."

Pascal is from Paris. He speaks French and learns English. He thinks I'm also somewhat French because we both have dark hair. He likes to play with me. His parents work at the United Nations in Geneva. They often travel to Africa. Pascal plays a violin that he has brought with him. He takes private lessons from a teacher close by. I have almost forgotten how to play piano. It was long ago that I last touched my Bechstein. Monsieur Bagnoud told Papa that I could take lessons from a piano teacher who regularly comes for some other children. I'm not sure if I want it.

The week is over. Papa and Mama ask me if I want to stay at the Chaperon Rouge. I say yes. They don't want to make the same mistake as at the North Sea. I know they are leaving. Knowing makes it easier. They will call me every week and write to me. In two and a half months, we'll have Christmas vacation. I'll fly from Geneva to Düsseldorf, all by myself, and Papa will pick me up at the airport.

But everything looks different with Papa and Mama gone. I liked it as long as they were here, and I could talk and laugh with them. I liked Monsieur Bagnoud, Monique, Pascal, Miss Hischier, and many children. I liked my room, the view over the valley and the high mountains on the horizon.... But now, without Papa and Mama, everything looks different.

I don't mean things have changed. No, everything is the same, cheerful and friendly as before—but different. I walk the narrow path to the inn and enter from the back door to see if Mama and Papa are really gone. Mrs. Dick confirms that they left very early in the morning, because Papa has to be in the institute to teach tomorrow. Nonetheless I wish in my heart that they were still here. I come out from the inn through the front door. Yes, Papa's car is gone. I walk back to the school. Monique has already been looking for me. She seems happy and relieved.

The following day after lunch, Mama calls me. I am in the bathroom, so she waits for me and hears voices of children in the corridor where the telephone hangs on the wall. She hears running steps and laughter in the background. By the time I finally come to pick up the receiver, she must have been worried without knowing what was going on.

"Minochan, Minochan, how is everything? Are you doing all right?"

Without being able to see her, I can't say much. I feel strange. So I just say "Fine..."

"You sound different. Something happened?"

"No...."

"I was waiting for a long time on the phone. Didn't Monique tell you?"

"I was on the toilet. Big one. That's why."

Mama bursts into laughter. Her laughter echoes through the corridor. I feel better. Yes, now she is close. I tell her that Miss Hischier wants to learn the Japanese alphabet from me. I showed her already *a, i, u, e, o...ka, ki, ku, ke, ko, sa, shi, su, se, so*. Each time, she says, oh lovely, oh pretty, how fascinating, as if she were watching butterflies flying. So far, she has learned *a* and *i*.

Mama laughs again: "Well, Minochan, you have become a teacher now."

"And I can spell *Mino* in the Western alphabet."

"You are doing really well. And are you eating enough?"

"I get second helpings, most of the time."

Some children in the hallway are so loud. Mama asks me something I can't understand. So I'm quiet.

"Minochan, Minochan, are you there?"

"Yes." I'm getting tired holding the receiver to my ear. It gets heavy after a while. I can't see Mama anyhow. The school bell rings. Classes start.

I tell Mama I must go. She says she has been thinking of me all the time. She has been talking with Papa about me. I want to jump into her arms. I don't know what to say. So I only nod and hang up.

On a cool late-autumn morning, the entire school goes into the pine groves uphill. We are all dressed in our school uniforms. Gray jacket with light blue collar. Mama's embroidery is inside. I can feel it. I have been in the Chaperon Rouge for almost two months.

The children walk in pairs. I'm with Pascal in the third row. Monsieur Bagnoud walks at the end, giving directions with his big voice. Monique says he is a colonel in the Swiss Army. He is not tall but he's strong. He can easily carry two boys and run uphill. He is stronger than Papa. Miss Hischier walks in front. She has learned by now Japanese *a, i, u, e, o*. Still a long way to go before knowing all fifty letters. I have patience.

Monsieur Bagnoud calls out from the back that we are soon coming to a clearing. There we will have our lunch. All of us are shouting "Cheers!" "Hurray!" and "Hourraa-aa!!" I understand everything that Bagnoud says in French because he always repeats and pronounces it slowly. I understand what Monique is saying. I don't like our French teacher Monsieur Méthivier, because he is short-tempered and scolds everyone whose pronunciation is not yet quite right. We are soon going to tell Monsieur Bagnoud how we feel about Monsieur Méthivier.

At the clearing, we sit on logs and eat prosciutto, cheese, and tomato sandwiches. We can choose mineral water, milk, or orange juice. I think of Mama and Papa and what they are now doing in Cologne.

Mama writes in her letters that Papa is mending the back fence of our garden so that people can no longer stand there, gawking at how we eat with chopsticks. Sometimes people stand there through our entire meal. "Our house is not a zoo," Mama says. She draws the long white lace curtain, but we all would prefer looking over our garden with trees, including a large Japanese cherry tree.

Mama has found out that some of our neighbors are in fact nice people and interesting too—one trumpet player in the Cologne City Orchestra, one professor and head of a clinic at the School of Medicine, one bank director, one nephew of the late Chancellor Konrad Adenauer, one painter, and one architect. When I'm home for Christmas, Mama wants to invite them for a Japanese dinner.

Monsieur Bagnoud gives the signal to get up and walk back. We take a different trail. Tomorrow is Saturday. The morning is reserved for writing letters home. I want to tell Mama many things I see and hear, but I don't know how to write down everything in a way it happened. I start "Dear Mama, I want to see you and Papa." Then I stop and think. I want to write about Sundays. We can go to church if we want. There are two churches, one Catholic, one Protestant. I go to both. If it's a warm day, I go to the Protestant church, because it is cool. The pastor often sneezes. Mama said I should keep away from anybody sneezing. So I sit in the back, listen to the choir, but most of the time I'm thinking of how Papa is doing in his institute in Cologne. It's not too far from home driving along the Aachener street, wide and straight with four traffic lanes, toward the old center of town. The streetcars run in the middle. In his lab Papa once showed me many colorful crystals. He told me we are going to do experiments together. When I come home he will take me there again.

On cold Sundays I go to the Catholic church. The heating is better and the music sounds better too. Pascal always goes to the Catholic church. He tells me Dear Lord has the son, Jesus. And Jesus has a Mama whose name is Maria. So, Dear Lord and Maria are married, but Pascal says, no, they are not married, but he isn't sure. He also has a personal saint and worries why I don't have one. "You need a saint to ask for favors." Apparently, Pascal's saint does a lot of good things for him.

The statues on pedestals in the Catholic church and the paintings on the walls interest me. Some look as if they are alive. I can study them while listening to the organ music. Many painted figures have wings. Pascal says they are angels. They live in heaven and fly down to Earth whenever Dear Lord wants to tell us something important.

"Do they come down during the day or while we are sleeping at night?"

Pascal doesn't know.

"But they have trumpets. You must have heard those trumpets."

He is not sure.

"Have you seen them flying in Paris?"

He says no.

"Have you seen them here in Crans, flying over the mountains or through the forest?"

Pascal says no.

I'm going to ask Mama.

Sometimes I want to tell Mama about the night sky and how dark and full of stars it is over here in the Alps. Though we are supposed to be in bed by ten, I sneak out of my room, go to the playground, and look at the sky. I've never seen so many stars, not in Göttingen, not in Cologne. There are hundreds, thousands, millions of stars up there and some are very bright. Papa once told me about the Milky Way. That must be the Milky Way.

I stretch my arms toward the sky and wish Mama could have been here to see it.

―――――――

When I receive a letter from Mama, I put it in my drawer. A few letters from Papa, always typewritten and long. Mama tells me she also keeps my letters in the special box at home. I'm taking classes in German and in French. I'm learning arithmetic and calculus. Everyday life is in French, I argue with our French teacher, and I protest in French against Monique who once slapped me because I had eaten only half of my plate at lunchtime.

Monique doesn't understand that I could not eat that day because I was too sad to eat. I missed Mama too much. I want to tell her how much I enjoy skating and skiing, because it's winter now. Deep snow and the pond is frozen over. I want to tell her about a film showing a gorilla family in Africa. The mama gorilla carrying her baby on her back. I want to tell her about other films showing dolphins in the Pacific and colorful birds in the Amazon. It always happens suddenly when I become sad. In the middle of the day during class or during lunchtime. Nothing helps. I can't eat. I can't cry, I can't talk. I just feel frozen. It's very unfair that Monique slapped me.

One of the boys at Chaperon, Urs Nägerli, tells me he doesn't want to go home. He is the oldest boy in the school and the tallest. His family is in Zürich, but he doesn't have a mama. She left with some other man, and now Urs's father has a new wife. Urs says his father married her when he was three. Now he has two stepbrothers and another is coming soon. His father has a yacht in the Lake of Zürich and a cruise ship somewhere in the Mediterranean. Urs doesn't want to go home.

During vacation, however, Urs lives on his father's yacht and some employees cook for him. His father joins him on weekends, and when the weather allows, they sail over the lake. Sometimes they go cruising on the Mediterranean.

Urs is now eleven, and he knows everything about yachting, cruising, and nautical equipment. He has been in the École Chaperon Rouge for a long time. He's chewing gum most of the time, but you don't see it. He can do it very well. He gives me a stick of chewing gum during math class. I've never had one before. Immediately the teacher scolds me and makes me write seven times in German and French: "I'm not allowed to chew gum during class."

Because Urs talks so much about his father's yacht, Miss Hischier gives him a book about the history of galleons. I take a look. Marvelous photos and drawings. Urs lets me have the book until I finish reading it.

The book says the most important tool for seafaring is

the compass. The Chinese invented it back in the eleventh century. I think of Dr. Needham at Cambridge, who wrote so much about Chinese inventions. Surely, he can tell me about the compass. Urs always has his own compass in his pocket. Once he dropped it on the concrete sidewalk, but it did not break. The book also mentions how dangerous the sea voyages were in the sixteenth and seventeenth centuries. Often half the crews, sometimes three fourths, died because of poor food, poor sanitation or no medicine. The best medical treatment at that time was blood-letting and praying to Dear Lord. I feel sorry for the sailors. Urs invites me to come with him to his father's yacht during spring break but I said no, because I want to be with Mama and Papa in Cologne.

Mama takes me by streetcar to the old town center where the Cologne Gothic Cathedral attracts many tourists. We visit the Praetorium, the palace, where the Roman governor lived almost 2000 years ago, when "Cologne" was "Colonia" and was the regional capital of the Roman Empire. For centuries nobody knew that the palace was there, because it had been overbuilt in the Middle Ages. During World War II most buildings were destroyed, including the one above the Praetorium. During clean-up, the diggers found the old ruins. Now it's a museum.

Mama and I look at the mosaics covering the floor and many walls. This was the palace where Nero's mother was born and grew up, but many people don't like Emperor Nero. We look at sculptures, ceramics, and glasses on display. How good that they found and excavated the Praetorium. You never know what will pop up at such an old, old place.

The same is true for the Cologne Cathedral, just a block from the Praetorium. Here too they discovered the foundations of a temple dedicated to the highest god of the Romans,

Jupiter, and beneath the Jupiter temple the ruins of an even earlier holy site for the Germanic gods.

I wonder whether there were angels at the Jupiter temple and the Germanic holy site. Mama doesn't know but tells me that angels are said to be messengers of Yahwe, the god of the Jews, of the Christians, and of the Muslims, who call him Allah. Mama says the Renaissance painters created angels with wings, often naked and with childlike faces, some with a trumpet. Grown-up angels like Michael, Gabriel, and Raphael have clothes on and the biggest wings. Mama says angels are spiritual beings. If you believe in them, they exist. If you believe that they fly, they are flying. Well then, Pascal doesn't really believe in angels.

We walk to the Peking restaurant opposite to the Cathedral, and Papa joins us for lunch. Afterwards I go with him to his institute. We want to prepare for an experiment during my summer vacation. Papa has graduate students who work during the summer. I'm going to meet them.

So, the Spring vacation goes by and now the summer vacation is near. Papa will come by train to Chaperon Rouge, twelve hours straight. My suitcase will be ready.

MORE SCHOOL

THE HELLENISTIC WORLD

I'm back in Cologne for good. I am turning ten and ready to go to high school for the next nine years. The high school we have chosen is Kreuzgasse, the oldest in Cologne, going back to the twelve hundreds. It has its own campus in the Green Belt, conveniently located close to the university but on the other side of the Aachener Street.

The Kreuzgasse is one of four public high schools in Germany, fully bilingual with half the teachers German and the other half French. Kreuzgasse promises that, at the end of nine years, I can graduate with both the German abitur and the French baccalaureate. These certificates open the doors to all German and French universities. Now I'm settled in this school and I like it.

Every morning when Papa drives to his institute he drops me off at the Kreuzgasse. I can easily get home by street-car, though most of the time it's rather crowded at one P.M., because other schools also end around the same time.

I take Mathematics, Physics, German, English, and History, which are taught in German. By contrast, Chemistry, Biology, Geography, and of course French are taught in French. No classes in music, art, or religion are offered. One afternoon per week is for sports—swimming, soccer, 100-meter dash, long jump, high jump. I have to participate in just one of them, but no scores are given. Nobody really cares if I just hang around with Roland in the corner of the field. Roland and I chat about art and the newest exhibition in town. He is a real art buff. He wants to become a museum curator. We joke about it a lot while our classmates are doing their 100-meter dashes.

I enjoy Math, Science, and History. The History teacher started with the Greeks and Romans but his voice is monotonous except when he has to pronounce Greek and Latin words. I'd like to learn more about history. Mama says I should check out the city library. It surely carries many books on the Greeks and Romans. Ancient Egyptians too.

My textbooks in German, French, and English are easy to understand, but some teachers make them sound complicated by dissecting and reassembling the sentences in a way that in the end they barely make sense. I'm bored most of the time. I think of the guinea pig community—of Taro, John, Bob, Antoinette, Ali, Philippe, Odette, Karl, Jürgen, and Colette and how they are doing. They live in our garden under the cedar tree, housed in a several connected cages built by our handyman. They chat with each other incessantly, while eating grass all day long, keeping the lawn short.

When I bring vegetable leftovers or fruits, you can see who the boss of the guinea pigs is—Taro, the Patriarch, taut and strong willed. He jumps on the food and munches it as long as he wants. Others have to wait. With brown and white fur, Taro is good looking and confident of his position. In the beginning, we had just him and Antoinette in one small cage. Within a few weeks they produced little ones, who in turn multiplied. Our Guinea pig community reminds me of Chaperon Rouge, where children with many different personalities live under the same roof. They chat, eat, and sleep together. Guinea pigs don't learn anything in class, no physics or history. They are happy eating and chatting all

day and sleeping all night. Mama says she sometimes wants
to join them.

After school, students disperse. On my way home by
streetcar I get off at the city library and look through the
shelves for books on the Greeks and Romans. I check out six
books. I jump on the next streetcar and go home.

After lunch with Mama I hurry up to my room and start
reading one of the books on the ancient Greek world. After
the death of Alexander the Great in 323 B.C., the huge territories he had conquered became three kingdoms. The first
was Egypt and parts of the Middle East, the second was Syria
and the remains of the Persian Empire, and the third was
Macedonia, Greece, and most of Asia Minor.

Egypt was the wealthiest of all three, governed by Ptol-
emy and his sons. They built the Library of Alexandria, the
greatest in the ancient world, with maybe up to a million
papyrus rolls from the time of the great pharaohs, from the
Hellenistic and pre-hellenistic world, from the Phoenicians,
and even the Sumer. Alongside the magnificent library stood
the museum, the shrine dedicated to the Muses, a center of
intellectual discourse. Famous philosophers and mathematicians gathered there to discuss their ideas. I go through one
of the other books I borrowed from the city library; it had
a list of the philosophers and mathematicians with descriptions of their main achievements. I'm so intensely imagining
them sitting together in Alexandria that I can almost hear
their voices.

One of the scholars who certainly participated in the
debates is the great astronomer Aristarchus of Samos, who
lived in the third century B.C. He is probably the one who
came up with the heliocentric theory. He said the sun is the
center and stands still, while the Earth and all other planets
revolve around it. Aristarchus is supposed to have written
extensively on astronomy, but all that survives is a short essay

on the sun and the phases of the moon and estimates of their distances from the Earth. He used mathematical calculations and presented his theory as a hypothesis.

I'm astonished to read this, because Aristarchus had no instrument to help him observing the nightly sky. Just his eyes. Looking up at the Milky Way so often at the Chaperon, I know how difficult it is to find the stars or to measure the distance from the Earth to the moon. It is just impossible.

Aristarchus's ideas were dismissed by other philosophers and astronomers of his time, who believed that the Earth does not move, while the sun's daily procession over the sky from east to west can't be disputed.

So, Aristarchus' heliocentric theory was ignored and forgotten until Copernicus rediscovered it more than one thousand five hundred years later. Copernicus reached the conclusion that the sun sits in the center, and the Earth and the planets circle around it. In 1543, only after his death, his book *The Revolution of the Heavenly Spheres* was published.

Why only after his death? The book says Copernicus feared the Inquisition of the Catholic Church.

Why the Catholic Church?

I'm confused. The priest in the Catholic Church in Crans-sur-Sierre, where I went so often on cold Sundays, only talked about God and Jesus, and about angels and saints. Never about science.

At dinner, Papa says, "At Copernicus's time the Church was different. The Church considered itself the keeper of the word of God. As the keeper of the truth."

Mama agrees. "In the Old Testament you can read at three places that the Earth stands still, since God fixed the Earth on its foundation."

"Exactly," Papa chimes in. "For the Church the Earth was the center, and it didn't move. God created humans to conquer the Earth and all living things, but then Copernicus comes and writes that the Earth is not the center. The Earth is moving around the sun. That's against the Bible, against God. That's what the Church didn't like."

Why all this fuss? The Bible is not a science book. The Church made a mistake. Everybody makes mistakes. So, admit it and move on. The priest in the Catholic church in Crans-sur-Sierre told us, if you repent for what you have done, God forgives you. Then all is fine. The Church at Copernicus's time should have known that.

THE BULLY

I'm standing in the noisy classroom during a break between two classes. I'm standing by myself at the window thinking about Aristarchus. I'm wondering how many years he must have watched the sun and moon and how he felt when he realized that Earth was moving.

What a pity that so many manuscripts of the ancient world were lost when early Christians ransacked the great Library of Alexandria, the largest collection of papyri of all times, an estimated million rolls, brought together over many centuries. Just imagine how much knowledge had been accumulated. For the early Christians, however, all this was nothing but pagan filth and grime. They murdered everyone in the library and set it ablaze. They made sure that not a single papyrus survived. Books by Aristarchus were surely among those lost forever.

Suddenly somebody grabs me from behind and strangles me by the throat. I manage to wiggle out and turn. It's Thomas Wasserberg, the bully in our class. Before I can get a stable foothold to defend myself, he kicks me in the stomach. I fall to the floor and black out. I hear voices as if from far away. "Thomas knocked Mino down."..."Mino passed out."

I roll over and manage to get up. I grab my satchel and hit the bully over his head. In this moment Mrs. Bauer must have come into the class room. She screams at me. "Mino! What are you doing? Stop it!" She pulls me away. "You should be ashamed of yourself."

"Thomas grabbed me from the back. I was defending myself."

"Don't make up things," Mrs. Bauer screams, as if her hair were on fire. "I saw Thomas standing and doing nothing. You hit him over the head."

"Yes, because he had choked me and then kicked me in the stomach."

"Don't make up things. I saw you beating Thomas. I should have brought my camera." Mrs. Bauer leans forward and shakes me, a stout woman with bright red lips, making threatening gestures. "Don't lie."

"I'm not lying."

"I saw it. Thomas was standing there, doing nothing, and you hit him with your satchel. Don't make up stories."

"I defended myself. Thomas grabbed me from the back and choked me. Then he kicked me in the stomach so hard that I fell and blacked out. I have a right to fight back."

I see a malicious grin on Mrs. Bauer's face. Her voice becomes high-pitched. "Don't lie. I saw with my own eyes what you did to Thomas."

"Shut up, you monkey!" I yell at her at the top of my lungs. "You didn't see how it started."

"Get out, Mino," Mrs. Bauer retorts, pointing at the door. "Get out."

"Why should I get out? I've done nothing wrong but fighting back. I've a right to defend myself. Stop calling me a liar, you monkey."

"Get out of here—"

"I am not leaving." I point my finger at her face. "You get out."

Suddenly Mrs. Bauer quiets down. She starts the class, as if nothing had happened.

After class I go to her and say, "Mrs. Bauer, I've never lied in my life, and I've never been accused of lying before today."

Without a word, she walks to the door, but then turns around and gives me a nod.

One of the girls comes to me and whispers, "You're right. Thomas is a bully. He started it."

Another girl approaches and adds, "We always stay clear of that guy."

"You defended yourself. That's right. You're okay."

I thought of the Chaperon where many children lived together, twenty-four hours a day. Monsieur Bagnoud told us to defend ourselves when attacked. I had gotten into a few fights, nothing serious. Nobody attacked me from behind, but I clearly remember what Monsieur Bagnoud said and how he said it: Defend yourself whenever you're attacked. The Swiss are proud people. They are ready to strike back. That's how they avoided being drawn into World War I and World War II, when whole of Europe was on fire. When Monique slapped me on the cheek for not finishing the meal, I went to Monsieur Bagnoud right away. He called Monique to hear her side of the story and then scolded her for slapping me. She never did it again.

The classes have ended. I take the streetcar in the other direction from my way home. I get off at the main train station near the Cathedral, and walk through the tunnel to the other side, where the Cologne Music Conservatory is. I'm going to sit in my first class in the history of music.

It all started during a conversation with Mr. Gerhardt, my math teacher. He plays violin. I told him about my interest in the history of music. Many melodies come to my mind as I walk through the park or sit in the streetcar.

Mr. Gerhardt advised me to go to the Music Conservatory

and inquire whether I could be a sit-in student for some afternoon or evening classes.

Two weeks later the Conservatory called me in for a test. I had to play piano for five minutes and write a one-page essay. I passed and got a registration card as a guest student, which entitles me to sit in classes and attend evening concerts.

Before history of music class starts, I open my lunchbox. Mama made three big *onigiri* with rice, eggs, dried bonito, salted plum, and *nori*. She had given me one Deutschmark to buy Apollinaris, the bubbly mineral water I love to drink.

"Crete was part of the Greek civilization," I hear the professor say at the podium. "Cretean songs have survived, written on papyrus or engraved on stone. They mention gods like Helios, Nemesis, and the Muses." An image of a papyrus appears on the projector screen in front of the classroom.

I sit in the back of the auditorium but cannot see very much. All these other students in front of me are so big. I can hardly see anything. I get up and walk to the front row. I'm twelve years old, but with a guest student card, which I carry with my library card and the season ticket for the streetcar.

"Today we can decipher these songs with a certain degree of accuracy. Melodies were considered important in classical times, more than rhythms." The professor goes on talking about modes and harmony. I am disappointed that he doesn't play any songs dedicated to the gods of Crete.

I remember one evening at home when Mama talked about Greek music. She had just come back from Greece, where Dimitrios Rondiris had invited her to a series of classical tragedies performed by his Piraikon theater company in Athens. To recreate the chorus songs of classical plays, Rondiris had recorded folk songs on islands in the Aegean Sea. He had identified basic melodic themes, Mama said, he thought must be similar to chants that had been sung during the plays by Sophocles and Euripides written two and a half thousand years ago.

In the next class, one week later, the professor talks about the Romans and how little they added to the Greek musical tradition. To worship their gods the Romans played drums, trumpets, harps, lyres, flutes, and lutes. After Rome became Christian, the new Church banned all musical instruments. They were declared the Devil's work. Only songs by monks and priests were allowed. Under the sixth-century Pope Gregorius the Church regulated even those and they later became known as Gregorian chants.

I know what Gregorian chants are. Mama has one LP record. She played it for me. All songs sounded solemn and grave. Boring. Not surprising, as I learn in the Music Conservatory class, that, in the Middle Ages, the people began to yearn for other music—anything but Gregorian chants.

According to legend, in a village near Cologne, peasants were seduced by the Devil to dance and sing. They had songs and dancing tunes, but the Church declared them to be of the Devil. As punishment half of the peasants were turned into trees so that they couldn't move for an entire year. The other half had to dance non-stop for the entire year, until the Archbishop of Cologne freed them from the Devil's spell. But the ban on music continued.

The church softened its stance only at the height of the Middle Ages, when the organ was introduced as an accompanying instrument for men's choirs in churches and monasteries. I was thrilled to learn that the pipe organ had been invented in third-century B.C. in Alexandria. When Aristarchus came to Alexandria to present his heliocentric theory at the library, he might have met the engineer Ctesibius, the inventor of the pipe organ.

The professor points out that the organ was further developed in the second century A.D. and used in Roman theaters and festivals. The really exciting development in music, however, came only when towns and cities in Europe grew in size, as wealth was accumulated in the hands of merchants.

It was towards the end of the Middle Ages and the beginning of the Renaissance. Rich merchants started to compete with the Church, paying composers and musicians. In return, the composers and musicians created music that reflected the taste of their paying patrons.

Not to be left behind, the Church in turn began to follow the popular taste, so much so that the distinction between sacred and profane music became blurred. The Church even started to integrate popular melodies and rhythms into Sunday services, so that people would come to church.

At home Mama tells me that Taro, the alpha male of our guinea pig community, died on the lawn, with grass still in his mouth. During the past two weeks, Bobby, one of Taro's sons, had become aggressive towards his father. He started to push Taro away from the best food. Taro lost his position as the alpha male and was ousted from the group. I saw him during the past ten days mostly alone, separated from the others. Though Taro continued to look handsome and agile, he tried to stay as far away from Bobby as possible. Deep inside he must have suffered from his defeat. Taro died of a heart attack.

We bury him on a bed of dandelion leaves under the cherry tree.

NATURE AND GODS

I've decided to call Mama "Saya," because Papa calls her Saya. It has been her nickname since schooldays.

I now have my own passport. This means I'm almost grown up. I carry my own suitcase. It's summer vacation. Saya and I are flying to Japan. Papa stays in Cologne, but he told me before departure I should protect Saya, since I'm the man in the traveling party.

Grandpa's Shinto shrine, where Saya grew up, is really impressive. This is my third visit and each time the place looks more surprising. Twenty-five acres of cherry trees, oak, pine, cedar, chestnut trees on the top of a hill in the middle of Kyoto. On pictures taken from an airplane, the hill looks like a ship riding the waves over rows of houses.

One morning I walk in the forest, trying to catch cicadas with a butterfly net. They seem to sit everywhere on the trunks and branches. Saya told me that she used to go around and snatch one with a net, but let it fly again soon because she heard the cicadas would live only one week to ten days. The males sing all day to attract females and the females lay piles of eggs on the branches. The larvae fall from the branches to the ground and dig into the earth. There in the darkness of the underground they live as nymphs for seven years, eating roots of trees and bushes. After seven long years, they come out when the ground gets warm in the summer and start the cycle all over again.

I finally catch one cicada, which immediately stops singing. It murmurs between my fingers, kicking its tiny legs and trying to fly away with its transparent wings. "Don't be afraid, *semi-chan*. I admire your delicate wings and the musical instrument on your belly. I wish you a wonderful summer!"

I watch it fly away. I'm saddened by the idea that it is going to live for only a few more days. Why so short after having waited in the darkness of the soil for seven long years? At least it can now enjoy the warm air, can choose from hundreds of trees to sit on and feel sure in the company of countless other cicadas. It doesn't know how short its life is. What it must do now is make music by rubbing its legs against its belly. He must make his music as powerful and intense as possible, all day and every day until a female cicada is enchanted.

Standing among the trees and surrounded by the chorus of cicadas, I trace a certain rhythm, cadence, infusion and find myself in a strange stillness. I'm not listening any more. I disappear in the stillness of chorus.

"Minochan!" I hear Grandpa's voice. Wearing the white summer robe of a Shinto priest, Grandpa catches up with me and takes my hand like Saya does. He is almost as tall as Papa, but very slender. We walk the path together into the forest, which surrounds the Shinto shrine. We sit on a rock under the shadow of maple trees and listen to the cicadas. Here their music is different from before, because the trees are different. The music is in a different key. It sounds as if thousands of cicadas are pecking in unison.

"They sense there will be thunder and rain this afternoon," Grandpa says. "That's why they are hurrying along. When it rains, the cicadas stop singing and hide under the leaves."

"I want to hear thunder, *Kaminari-san*. When I was small, Mama told me about the Thunder god who lives in the sky. Now I know from my physics class that thunder is caused by electric discharges in the atmosphere, but I still think it's fascinating to imagine the Thunder god."

"Since ancient times in Japan," Grandpa says, "*Kaminari-san* has been the god of water and of rice. In China, he's depicted with drums over his shoulders, and I'm sure you know that, in ancient Greece, he was called Zeus, the chief god of the Pantheon, and in old Rome, he was Jupiter."

I never imagined Grandpa as a Shinto Priest would be so familiar with the gods of Olympus. Before the summer vacation, I was reading about Greek mythology, particularly about the twelve main gods and their roles and their qualities, good and bad. Hera, the wife of Zeus, was the highest goddess, responsible for a happy marriage and safe pregnancies. Poseidon, the god of the ocean, protected voyagers. Athena was goddess of wisdom, art, and strategy. Apollo, the Sun god, was also in charge of poetry, music, and prophesy. Aphrodite was the goddess of beauty and love. All these ancient gods were immortal, but not perfect. Hera, for instance, is known for her jealousy. Ares, the god of war, for his cruelty and moral indifference. The gods were sometimes mischievous or brutal. I like the story of Pandora, made by Hephaistos, the god of fire and blacksmithing. Pandora was the first human, a woman, and Hephaistos used just a lump of clay. Other gods and goddesses endowed her with charm and glamor, and sent her down to Earth. She was given a box but had to promise never to open it. Out of curiosity, however, she opened it. All evils of the world jumped out of the box before she could close it shut again. Only "hope" was left inside.

Grandpa seems to know all about the Pandora myth. So, I ask him whether the Shinto gods and goddesses also gather to talk about the problems of human world, like the gods and goddesses of the Greek Pantheon on Mount Olympus.

"Every year in October, the Shinto deities are thought to gather from all over Japan to meet at Izumo Taisha along the coast toward China and Korea. There they talk about the past year and what to do in the coming year. They stay for the whole month." Grandpa smiles and pokes my cheek just as Saya often does.

"We are lucky in Japan," he continues. "The Shinto deities are still alive because so many people think of them. They live in the forests, in the streams, in the waterfalls, in the mountains, at all places where people are able to still feel the *ki*, the energy of nature. Since ancient times, people in Japan have built shrines and temples. They make offerings to the gods and ask them to grant wishes. Out of these myths came rituals with prayers, music, and dance...joyous festivals. How fortunate we are that people still feel the presence of the gods all year around."

"Like in ancient Greece?" I ask.

"Yes," Grandpa replies after a long pause, "except that

the gods of ancient Europe have long been chased away.
They are no longer spiritually alive. Gods die when people
stop thinking of them, stop believing in them and caring for
them."

———

Grandpa tells the deputy priest that he would be away with
me for a few days. I pack my backpack for the trip. Early in
the morning, we walk down the long flights of stone stairs,
where young trainees are sweeping up the leaves. They bow
as we pass.

During several hours of train-ride, Grandpa buys two
lunch boxes and two bottles of tea. He gives me his grilled
salmon, steamed egg custard and cooked bamboo shoots,
which he knows I like. The train runs along the Pacific Ocean
and I ask Grandpa where we are heading. I have been curious since we began our trip, but he keeps it a secret until we
reach our destination.

The train stops in a valley surrounded by steep mountains. From there we take a bus to where the road ends and
walk into the forest. High cypress, cedar, and pine trees. They
are so dense that it's almost twilight where we walk. The tree
trunks are three times wider than I can span with my arms.
From somewhere deep within the forest, I hear the sound
of rushing water. Then, a few steps farther, I see a waterfall
thundering down from very high, magnificent and white,
splashing, rolling, and howling. I can't make out where it
begins.

"More than one hundred thirty meters high," Grandpa
answers, guessing my question.

"Higher than the towers of Cologne Cathedral—much
higher."

We get closer. Looking up into the veil of water, I feel
dizzy.

The roar of the waterfall fills the air. Grandpa tells me

in a chatty way that in centuries past, people came from far to see the waterfall. He takes me deeper into the forest. We walk for quite a distance along the valley floor. We come up to a majestic cypress tree.

"I have been here before," he says in a low voice, so as not to disturb the silence that envelops us. Keeping his hands stretched out toward the tree, Grandpa closes his eyes. I do the same. I feel warmth streaming into the palms of my hands. I look at Grandpa. He stands next to me with his eyes closed.

On our way out of the forest, he tells me that he visited this tree more than forty years ago when he was still a young priest. The tree told him that he could come back any time, listen to the silence, and feel the energy of nature.

In the evening Grandpa takes me to an inn on the Pacific shore, where the water from a hot spring gushes day and night into a large pool next to the ocean. We sit in a corner of the pool, in the soothingly warm water. Several men are sitting here and there, lost in thoughts. One is singing. More and more stars are filling the sky.

"Minochan, you are already twelve, growing up fast," Grandpa says. "I still remember the time when you were a baby. I came to see you and your parents in America. On the last evening before leaving for Europe, I carried you into the night and we looked at the stars. You were gazing at the Milky Way." Grandpa starts to hum a melody I know so well—one of Saya's songs when she used to put me to bed.

Next morning breakfast is served in our hotel room. It's a banquet with many different dishes. Saya should be here. At home in Cologne, she often tells me about hot springs and the breakfasts they serve in the hot spring inns. But Saya is in Tokyo right now talking with people for a TV production for German TV. At least I'm here with Grandpa eating grilled fish and fried oysters with different vegetables, miso soup, rice, seaweed, egg cake, and little bites of melon.

"We are fortunate in Japan," Grandpa says, "surrounded by the Pacific Ocean and the Sea of Japan, where warm and cold ocean currents meet. That's why there is such a richness of sea life along our shores.

"At the same time," Grandpa adds, "there are earthquakes, large and small. Volcanoes erupt and typhoons come—twenty, thirty typhoons—every autumn. Japan is a restless place. That's why the people living here have opened themselves to the power of nature. Nature nurtures but also destroys."

We finish our breakfast. We go out to the water's edge. Black rocks everywhere. Slippery algae. In one of the tidal ponds, I see a large fish, a red sea bass gasping for air. It is so large it can hardly move in the pool without hitting his head against the rocks. The tide is still receding.

Grandpa squats close to the pond and stretches his hands over the surface of the water. He calmly looks at the fish. It stops struggling. Grandpa nods to me. I grab the fish and lift it out of the water and walk over the rocks to put it back into open water. There, however, the fish stays at first very still at the same spot, then it starts swimming in tight circles, breaking the surface and jumping up several times. After a final jump, its glittering dark red body disappears into deeper water.

We continue our walk along the rocky shore. I keep thinking about the waterfall that we saw yesterday and tell Grandpa that Papa talked with Dr. Needham about waterfalls. Dr. Needham has written in one of his volumes on Science in China about Taoists who used to meditate close to waterfalls.

According to Needham, waterfalls cause the air to become negatively charged. Negative air ions relax the mind, while positive air ions do the opposite to all living things. Near any waterfall the negative air ions give people the feeling of relaxation and peace.

Grandpa nods. He tells me that Dr. Needham came to visit him in Kyoto and spent a full day with him talking about *I Ching,* the "Book of Changes," and many other aspects of ancient Chinese philosophy.

We sit down on a rock facing the ocean. Many small islands seem to be floating there and the sun is behind the clouds. Fishing boats return to the village harbor.

"Grandpa," I ask, "mountains, waterfalls, hot springs, ocean, fish, all sea creatures, all animals and humans, isn't everything created by God?"

Grandpa looks at me for a moment and smiles. "Growing up in Europe, you must have heard this story often enough." Then, serious again, he casts his eyes over the ocean. "No," he says after a while, "there is no such thing like a creation, one-time event that would produce something that can last forever. Nothing lasts forever. Nothing stays the same. Everything changes from day to day, from year to year, from century to century, and this has been going on as long as time exists. Everything evolves. Hot springs stop here and start there. Some volcanoes become dormant, others wake up. Mountains are worn down with time. Ocean floor turns into mountains. Even the stars are born and die. We humans share the same fate."

"But God...?"

"God or gods or nature—it's all the same by different names."

"But there is only one God, the Almighty God. Isn't this true?"

Grandpa doesn't answer. He remains silent. "Minochan," he then says while still gazing into the distance, "I know you've never been to a desert. Nor have I, but I've thought about the people who spend their lives in a desert, any desert, generation after generation. Nothing but sand as far as the eye can see. A deadly hot sun during the day and chilling cold nights. To find water, you have to find an oasis, days, maybe weeks away on foot or by camel. Remember, the people who wrote the Bible were people of the desert. They were different from most of humanity. For them nature was not nurturing. Nature was hostile, dangerous, malevolent. These people dreamed of being able to control nature. That's why they wanted to have a god that is above nature. Moses was the first one to declare his god to be such a god, the only God, to be in control of everything—nature and humanity. He declared his god to be the only god, the only true god. A god who demands to be worshipped at the exclusion of any other deity."

"The Almighty God?"

Grandpa nods. "The seed of many great thoughts and achievements, but also the seed of intolerance, fanaticism, persecution...."

The tide is returning and the waves that break on the rocks are bringing water back into the tidal pools. We feel the splash of the waves, and spray droplets wet our skin. Grandpa stands up and takes my hand.

On the train back to Kyoto, Grandpa tells me the story of Giordano Bruno, a Dominican monk, philosopher, and scholar during the Renaissance. Influenced by the Copernican heliocentric theory, Giordano Bruno became convinced that space must be immense, that there must be countless stars beyond the stars we can see. Infinite and boundless. He wrote that, if God exists, He is too great for humans to grasp. God is beyond definition.

Giordano Bruno refused to believe in the God of the Bible as preached by the Church. He said God is much larger. Giordano Bruno was declared a heretic and burned alive at the stake in Rome in 1600.

I'm thrilled to hear Grandpa talk about Copernicus and his heliocentric theory, but I'm angry to learn what happened to Giordano Bruno. It's true what he said about stars and about the universe. Looking at the night sky, I too feel that there must be many more stars beyond the ones we can see. There are galaxies beyond the Milky Way—many many, galaxies with countless stars. What's wrong with saying so? Why did Giordano Bruno have to die?

How fortunate it is that the Church isn't doing such atrocious things anymore.

COMPUTER LORE

Papa hates computers. In Cologne, when he wants to solve a crystal structure, he has to punch holes into some five thousand punch cards. He showed me the machine, like a big typewriter. He has to bring his punch cards to the Computer Center, the only place at the university with a computer. He has to hand over his tray with five thousand punch cards and is told to come back in a week. One week later Papa gets his tray back but without any results because, as the woman at the front desk tells him, there is one error in his punch cards.

To find this one error, Papa has to check every hole he punched in his five thousand punch cards. Of course, that gave him a pretty bad mood, but times are changing. Papa tells me that two assistant professors at the University of Cologne, one from Physics and one from Math, are gathering a group of graduate students to form a Computer Club.

"May I join?"

"Of course." Papa takes me to the next meeting. I'm the only school kid, fourteen years old, among the graduate students and assistant professors. That's how I got an early start into the world of computers.

I learn BASIC. I learn how to write computer programs in machine language. The only big problem is that we have no computer that I could use. The Math Department has one and the Physics professors are talking about buying one for their department. I think I should build one for myself, a small one. Papa agrees, though he looks a bit puzzled, wondering how I would be able to pull it off. By this time, I've learned so much at our Computer Club. I feel I can do it.

My math teacher at the Kreuzgasse, Math-Gerhardt, also joins the club. He helps me to sign up for the *Computer Magazine*, which I now receive every month. It comes directly from California. Papa pays the subscription fee. I find ads for different components that I would need to build my own computer. They are on sale but still very expensive. Besides, they have to be shipped all the way from the USA.

While I'm pondering what to do, Saya returns from the ARD TV studio with a big check for her latest TV-documentary. She listens to my plan and gives me her check with a big smile, though she has no idea what I'm going to build.

I set up my own lab in the basement with a workbench, several boxes, drawers and an electrical tool-kit. It could take at least one month before the first shipment would arrive from California. I make a detailed plan how I would spend the following weeks so that I can concentrate on building a computer when the big package arrives at our doorstep.

Every morning I have the classes in the school, and two afternoons each week I spend in the Music Conservatory attending classes for the History of Music. By now we are dealing with Johann Sebastian Bach. It is fascinating to hear how his music embraces every musical genre of his time. He contributed by opening up new horizons, particularly counterpoint or contrapuntal passages. He composed the beautiful *Brandenburg Concertos* between his mid-twenties and mid-thirties. We listened to the first part in class. I'm going to ask Saya to buy the LPs with the complete concertos for my birthday.

How fortunate that later composers, in particular Mendelssohn, rediscovered Bach and recognized him as the greatest representative of the Baroque era. Otherwise, Bach's *Brandenburg Concertos* might be lying buried somewhere in the forgotten history of European music.

I'll definitely continue with the History of Music class. Every week I'll attend a concert in the evening by a faculty member or an advanced conservatory student. In addition, I get a student discount for the Cologne Concert Hall. I've already listened to the Berlin Symphony, the New York Phil, the Boston Phil. I have listened to Vladimir Ashkenazy with the London Symphony, and the La Salle String Quartet. I particularly like to listen to soloists like Yehudi Menuhin, Andrés Segovia, Emil Gilels, Van Cliburn, Claudio Arau, Alfred Brendel, and Daniel Barenboim. The two last pianists were both in Cologne for a whole semester teaching a master class. The list goes on and on…. I must not forget Jasha Heifetz, Yo Yo Ma, Mstislav Rostropovich, and Maurizio Pollini, whom I heard four times. I've been around so many great performers that I decided to become a listener.

Sometimes I go with Saya to a concert, but mostly I go alone, because she is busy either proofreading her next book or touring somewhere in Germany, Switzerland, or Austria. I don't want to take Papa with me to the concerts. He can't distinguish Mozart from Händel or Ravel from Albeniz. He sings or hums off tune and once passionately applauded at the wrong moment.

I have not been doing my homework for school as well as I should. Math and all sciences are a great fun, History, English and French are all right; but German is the problem. Not the language but the teacher. I don't get along with him. He is short-tempered and easily irritated, and he doesn't allow any contradiction. Often he does not even allow a question. I don't like him and therefore I'm not interested in German. Saya once read to me a poem by Rainer Maria Rilke, "Panther." It sounded so majestic and sad at the same time that I wanted to read his other poems and I did. Nevertheless, I'm still bored in German class. As a result, I get a D. That's one grade better than an F, for "flunk."

One day one of the boys brought a copy of *Playboy* to the German class and sneaked it under the table to his friends. It came past me, and I looked at some of the naked women. All had big breasts, and they wore furious looks or half-open mouths. None of them smiled. All seemed terribly bored.

Every two weeks I return the books to the city library and take out six new ones. They keep me awake at night. I can't stop reading. So many things have happened in the world since the beginning of human civilization. I want to learn as much as I can. The Sumerians were the ones who developed writing, around 3,500 B.C. Needham writes in the book I took out from Papa's library that the Chinese invented writing at the same time, maybe even earlier. The country where the Sumerians lived was the Fertile Crescent, Mesopotamia, between the two rivers, Tigris and Euphrates. They built cities. They invented the wheel and the plough. They became masters in irrigation. Why did the Sumerian civilization collapse? What did the ancient Egyptians believe? What did they know about the underworld? What is written in Hammurabi's legal documents? Genghis Khan's Empire stretched from the Pacific to the Adriatic. How was it possible to rule such a vast territory? Why did Constantinople fall? Questions after questions. In the morning, instead of going to school, I want to continue reading and finish the next chapter.

Papa comes to wake me up. I tell him I'm sick. He puts his hands on my forehead and dismisses my claim. Saya always jumps up when she hears I'm not feeling well. Papa and Saya stand in my room next to my bed. I tell them I have a sore throat. Saya asks me to open my mouth. She looks and is instantly convinced that I do have a sore throat. She herself often suffers from it. Looking into my open mouth she says my throat is very red and swollen.

Papa asks her whether she has ever looked at my throat when I'm fine.

Saya is puzzled. "Why do I have to look at Mino's throat when it is not sore?"

"So that you can see the difference between a normal throat and a sore throat."

"When the throat is red, it's sore—"

"But the throat is always red," Papa interjects, "just a bit redder when it's sore."

"I often have a sore throat myself. I know the difference." Saya dismisses Papa's argument. Her vivid imagination works not only when she writes novels but also when she looks into my throat.

Papa still murmurs that a bit of sore throat is not a reason to skip school, but Saya disagrees violently, saying that Mino's sore throat could be the beginning of a serious cold or the flu. "He must keep warm and eat and drink a lot. He must get rid of his sore throat before going to school again."

Papa is still visibly not quite convinced, but he has no other argument. Anyhow he has to leave for the university. Students are waiting. I can stay in my bed. Saya brings a tray with a big breakfast. When she comes back and finds all plates empty, she is happy. I continue reading into the afternoon.

One day the expected big shipment from California arrives. I eagerly unpack everything and check all the components. I read the long instructions on how to assemble the computer. I have to put various components onto a board and solder many connections without getting too close to transistors with my hot soldering iron. I start doing it and test the connections. All are fine. However, I need an oscilloscope to see if I can produce the right signals.

After several inquiries in the Physics Department, Papa finds someone from whom I can borrow an oscilloscope.

Very useful. I'm spending afternoon and evening hours in my basement lab putting the system together. I learn how to write binary code. Finally, after the spring semester has passed and the summer vacation starts, I'm able to type a command into my keyboard and see the corresponding number on my screen.

My computer works.

I help Papa load it into his car. We bring my computer to the monthly meeting of the Computer Club. All of them—the graduate students, the assistant professors, and Math-Gerhardt from my high school—gaze at my creation, eyes wide open. All nod in synchrony when the numbers which I'm typing on my keyboard pop up on the oscilloscope screen.

"Well done," Math-Gerhardt says.

One week later, our whole Computer Club gets on a bus to visit the Radio Telescope on the Effelsberg, built and operated by the Max Planck Institute for Astrophysics in Bonn. From Cologne it's nearly a two-hour drive. With a diameter of 100 meters, the telescope dish is huge, reaching higher than the towers of the Cologne Cathedral.

We spend the whole afternoon and well into the evening in the control room, where astronomers are collecting data from faraway stars, especially stars that cannot be seen with optical telescopes. These stars are invisible, as one of the Radio Telescope staff astronomers explains, because they are obscured by those humongous dust clouds that mark the plane of our Milky Way galaxy. But the Radio Telescope can receive signals from these embedded stars. The staff astronomer points to one of the fuzzy smears on the large computer screen in the control room. That's where an embedded star is in the process of being born.

I'm overwhelmed to see the birth of a star. It makes me happy to learn that all these images could not be projected without computers. Of course, far more powerful computers than the one I've built.

"COME IN...SAY HELLO..."

We are spending the summer at Penn State, where Saya and Papa have met and where I was born. At that time Saya was doing her graduate studies in Theatre Arts. Dr. Walter H. Walters was the head of the department. Now he is the Dean of the College of Arts and Architecture. He and his wife, Gerrie, have reserved a two-room apartment for us, downtown, right next to campus. They brought in beds and bedding, a table and chairs, a sofa and lamps, kitchen utensils, towels, and even toilet paper, so that we can move in straight away and Saya can cook whenever we don't feel like eating out. It's a cozy place.

Saya has been excited ever since we learned we were going to visit Penn State. The first time after so many years. On our second day, Saya and Papa take me to the hospital in Bellefonte, a small nearby community where I was born in the midst of winter. We find the hospital boarded up and learn that it will be torn down in a few weeks. We can still go in. We walk through the semi-dark hallway, past room after room, which have signs like "Labor Room," "Delivery Room," "Baby Ward," and "Waiting Room." The air is sultry and stifling hot. Hard to imagine that my life began here, when this hallway was busy with nurses and doctors running around, and while there were five feet of snow outside.

Saya reminisces how it was at that time. The day after I was born, Saya says, the head nurse came with a syringe in hand to give her an injection.

"This will stop your breast milk."

"Why stop?"

"So that you can be free...don't have to nurse your baby...can go to parties or go shopping whenever you want."

"But I want to nurse my baby."

"No, no, this injection is good for you. Your breasts will stay in shape."

"But I want to nurse my baby."

The head nurse left. Shortly afterward the doctor came smiling. "You must understand: breast milk lacks certain ingredients—enzymes that are necessary for the health and well-being of the baby. Formula is much better. Better for the baby and for you. More nutritious for the baby and easy for you to prepare. All new mothers take the injection—it's just routine nowadays."

"But I want to nurse my baby."

Saya looks around in the dilapidated former delivery room. After steadfastly refusing the injection that would have stopped the flow of mother's milk, she was treated by the doctor and the hospital staff as a stubborn backward oriental woman, who could not grasp the blessings of modern Western medicine.

We drive to campus and look for the Graduate Circle, where Saya says we stayed for four months before moving to Europe. But we can't find the Graduate Circle. Everything is gone and replaced by new buildings.

We drive through town to look for the house of the Meteorology professor where Papa and Saya stayed before I was born. Saya tells me how many hours she spent there listening to classical music—Bach, Vivaldi, Haydn, Mozart, Beethoven, Scriabin, and many more. She tells me she walked every day from there to the Corner Room downtown, a one-hour walk, singing along the way. At that time there were mostly open fields. Now there are no open fields, only houses after houses, traffic lights, and lots of cars. Saya and Papa can't find the house of the meteorologist. We drive back downtown and Saya is happy to find at least the Corner Room, where Allen Street meets College Avenue. Still very crowded.

We walk onto campus and find the Schwab Auditorium, a heavy stone building with a broad staircase leading up to three arched entrance doors. Saya runs up the stairs and opens the middle door. She begins dancing through the foyer, up and down the stairs to the second floor. She comes down again, humming and beaming. She opens the double-winged door to the theater—a large hall with rows after rows of thickly padded red seats. About nine hundred seats. Saya runs down the center alley and up to the stage. She bows as if there were a full-house audience. Papa and I applaud.

Saya wants to show us Graduate Hall, where she stayed for two years. According to her, it was a pastoral wooden building, three stories high, dating back to the eighteen hundreds. Everything was made of solid oak—the entry hall, the staircase, the walls, the floors, the desks, chairs, cabinets, and beds—all smooth, warm brown oak. But there is no Graduate Hall. It's gone.

This evening Dr. Walters takes us to the Eisenhower Auditorium. He and Gerrie sit with us in the first row of the balcony. We are going to see the play *That Championship Season,* by Jason Miller. According to the program, the play was written in 1972, set in Scranton, Pennsylvania. It received the 1973 Pulitzer Prize for Drama.

The Eisenhower Auditorium is large, with two thousand five hundred seats. Saya is impressed by the size and the state-of-the-art technology. Had it not been for Dr. Walters' initiative and fundraising skills, it would not be here.

Saya chats with Dr. Walters. Papa tells Gerrie what we have seen so far. I'm sitting between them, still reading the plot of the play. I have never seen a performance of a modern play on stage.

In Cologne, I took Saya to Mozart's *The Magic Flute* because I was able to get two free tickets through the Music Conservatory. I liked Papageno and Papagena. Also Tamino and Pamina. Tamino is enchanted by the magic flute and Papageno by the chime of the magic bells. At the end Zarathustra unites the lovers. For days after the performance I could not get the music out of my head. Then I learned that Mozart composed this opera and conducted its first performance two months before his death. He was thirty-five. I agree with Beethoven, who said Mozart's *The Magic Flute* is his greatest work.

That Championship Season starts with two middle-aged men walking across the stage, drinking and behaving like teenagers. They incessantly talk, talk, talk with a lot of gestures. I can hardly follow. Three more men join them, also middle-aged. They say, "Jesus" and "Christ" a lot. They say, "I'm pissed off," "son of a bitch," and "kick his ass." They are certainly good actors, playing their parts with excellent timing. Everything on stage suggests that things will not go well. I look at Saya. She is not quite following either.

Next morning, while Saya and Papa are doing groceries, I go back on campus. As I walk around, I see the door to Dr. Walters's office open.

"Mino, come in…. Say hello." His voice has a particular color. It sounds as if the world is floating on the blue ocean. His voice is calming. "How did you like the play last night?"

"I didn't understand what they were talking about. Four middle-aged men and one older man drinking and talking and talking."

"I know, full of jargon."

"My mother didn't get it either. So, we are going to see the play once more tonight. She already bought tickets this morning."

"Good for you! You speak beautiful English. Do you read English books in school?"

"We have English textbooks, but they are dull. So, I read English books from my mother's bookshelf. I've finished

Tom Sawyer and *Huckleberry Finn*. Also Hemingway's *The Old Man and the Sea*. I like Robert Frost's poems, but don't understand many of them."

"You are reading more than our kids of your age," Dr. Walters says.

"And I've read George Orwell's *Animal Farm*. Twice. The first time, I was mad at Napoleon, the swine, who rules over the entire farm with the dogs as his police. The second time, I was mad at the sheep who let this all happen."

Dr. Walters nods. "I'm with you for the second reading. There will always be sheep. Can't help it." Changing the tone of his voice, he switches the topic. "You have a nice haircut."

We continue to chat until his secretary announces the visit of some other dean. "Come back anytime, Mino. I enjoy talking with you," Dr. Walters says, with a big smile and his melodious intonation.

During the rest of our stay at Penn State we walk on campus every day. We go to theater performances, not only in the Eisenhower Auditorium but also in the Center Stage, where the audience sits all around the stage. Years ago, it was part of the College of Agriculture and was used for pony shows. Once I go with Saya to the Pattie Library, while Papa is visiting his old lab in the Material Science building.

On the day before our departure, Saya stands in the middle of campus. "Many of the buildings I knew have disappeared. Most of my professors are gone. All my college friends are no more. Hundreds of strangers are walking around, as if the place belonged to them. What a queer world."

ARTIST'S TRIP

Back in Cologne, school has started. One of our neighbors across the street is a professor who teaches design and drawing at a technical school in the city. He is also an artist, a painter. Mr. Schwarzfeld is about sixty years old. A man of short stature and few words, but with mischievous eyes. His house is full of his paintings. They hang on the walls and are stored in the attic. From time to time he changes the display on the walls of the staircase and in his living room. He draws cityscapes and landscapes in Cologne and along the Rhine river or in small villages in the Eifel mountains or in Switzerland or Austria, where he spends his vacations. He is a master in many techniques, from ink drawings to watercolor to oil. His pictures are naturalistic, just as the scenes really are with a few interesting twists in the perspective and in the colors.

I ask him whether I can join him on his next drawing trip.

"If you have the patience to sit all day in one place."

"What kind of utensils and canvas do I need?"

"You should start with pencil drawings. This will make you sagacious for perspective. If you get the perspective wrong, forget it."

What Mr. Schwarzfeld says reminds me of a documentary I once saw with Saya and Papa. It showed Picasso sketching a scene naturalistically with black crayon, almost like a photograph. He then erased it and redrew it, now with fewer lines and suggestive forms. Though it looked to me so attractive, Picasso erased it again and redrew a third, a fourth, a fifth time. At the end the same scene had dissolved into lines, dots, circles and squares, which still retained the memory of the first sketch, but had become infinitely more absorbing.

If the foundation is solid, you can develop endlessly—that's what Mr. Schwarzfeld wanted to teach me, I think.

"Ready for our artists' trip to the other side of the Rhine?" Mr. Schwarzfeld calls from across the street. "Let's go."

We carry all our tools, two big sandwiches, and a few bottles of water. The streetcar takes us to the other side of the Rhine. Then we walk for fifteen minutes to reach our destination. Carrying an easel, a canvas, a folding chair, a paint box, and utensils for brushes, Mr. Schwarzfeld finds the spot where he likes to settle down. He plants his easel, places his canvas on it, and then quietly watches the riverscape for a long time as if to memorize the silhouettes, the lights and shadows.

I set up my easel a short distance away and choose a different direction, pine and oak trees dominating the front. Behind them, slightly to the left and rising above the trees, the old citadel from the Middle Ages reveals its imposing structure. Next to it stands a modern office building, all steel and glass. The leaves and branches of the trees are in constant movement. The citadel remains frozen in time.

I start to draw trees with light and dark shadings of many thin strokes. I try to capture their movement, particularly of the oak trees. The pine trees look heavier, like a dense, dark bundle, but they are also in flux. How can I draw a wind blowing through the leaves and branches? They are shifting and talking to each other. Then there is the tower in the medieval city wall. Intricate and well-balanced lines. How did the workers so long ago build such a complex structure? All stones had to be brought up by pulleys and set exactly in place. Just drawing heavy stone by stone I start to perspire, though the September air is rather chilly. I add the office building with square windows in faint straight lines.

When I am done, Mr. Schwarzfeld also stops and comes over.

"*Donnerwetter!*"

He looks impressed. He pulls out his medium-size pencil: "May I?"

I nod. Mr. Schwarzfeld adds shadows here and there.

"Mino, you've got it!" He taps my shoulder with a smile. We eat our sandwiches without a word. I hadn't realized how hungry I was.

We go out together two more weekends, again down to the bank of the Rhine river. I draw its fluid lines with the Cathedral on the opposite side as background. This time I don't use pencils but ink. I have two pens of different width, medium and wide, and the small square bottle of black ink that Saya brought from Japan. Most of the scenery I draw with the medium pen, adding shades and reflections off the water by spacing the lines differently. I use the wider pen only for the dense pine tree off to the side of my canvas.

At the end Mr. Schwarzfeld comes over and studies my drawing intensely. He finally nods. "You got it."

Saya invites Mr. Schwarzfeld and his wife for a Japanese dinner. *"Donnerwetter!"* he says "This Mino has a steady hand and a sure stroke. Ink drawings are unforgiving. There is no fudging, no erasing, no correcting."

After having learned from Mr. Schwarzfeld how to select scenes and draw them, I start to see landscapes and cityscapes differently. Some are suitable for tourist postcards, a bit stereotype. Only a few seem to invite me to draw them—from different angles, at different times of the day or of the year. Once I have drawn a spot, I remember it for good.

I also start to look with more subtlety at paintings and drawings done by famous artists. For instance, I admire J.M.W. Turner for his shades of color far more than before, and I admire Childe Hassam for his daring compositions.

One rainy weekend I go to the Wallraf-Richartz Museum downtown to draw figurines in the glass cases. They are miniatures—fighting, sitting, standing, dancing—whose movements caught my attention. These figurines remind me of Albrecht Dürer's charcoal drawings. They show people who can be strikingly ugly and unforgettably attractive at the same time. Saya has a book with all of Dürer's drawings. I've spent hours looking at them. On that weekend day, in the museum, I stand in front of Rembrandt, Rubens, Renoir, Van Gogh, and Klee. Imagining what they have seen and how they brought it onto canvas makes me almost dizzy.

In school, the fall semester began two weeks ago. I have joined the Advanced Math class, an extra-curricular activity taught by Math-Gerhardt. We are eight in the course, seven boys and one girl. We explore the possibilities of math, which seem to be infinite.

We also have a new French teacher, Mr. Winifred Bartenstein. He grew up close to Strasbourg on the border between France and Germany. He is fluent in both French and German, though his gestures are unmistakably French. He has a round face and sports a mustache. With his vigorous, dynamic manners he reminds me of Balzac. He has been at the Kreuzgasse for more than twenty years. He teaches upper division French language and literature. He is tough as a teacher, demanding. I've seen it. He asks question after question, but he also doesn't mind being interrupted any time. I'm looking forward to his class. He assigns at least one French novel per month. His enthusiasm for literature is contagious. I can visualize the characters of a novel just by listening to his recitation.

Mr. Bartenstein has already given us a list of novels we are going to read and discuss during the semester: Victor Hugo's *Notre-Dame de Paris* (1831) and *Les Misérables* (1862); Stendhal's *Le Rouge et le Noir* (1830); Honoré de Balzac's *La Comédie humaine* (1842–48), where Balzac tried to depict every level of the society of his time. Fifty volumes. Mr. Bartenstein is going to select one volume for us to read.

After nearly every class I stay to talk to him, since so many questions come to my mind just listening. He is endlessly patient and encourages me to think further on my way home.

FISH HEAD

Saya longs for sushi. After having been in Japan, I can understand how miserable she feels here in Germany. Here, in the fish stores, they call it fresh fish but what they have to offer are rather tired-looking slices of cod or plaice and smoked eel. There is herring in oil, in cream-sauce and with some other seasoning, but we can't eat herring every day.

We three once went to a Japanese restaurant in Düsseldorf, where fresh sashimi-grade fish are said to be flown in daily from Paris. Each sushi plate with eight pieces is way above Papa's pay grade as a university professor. We argued for some time whether or not we should walk out but finally decided to order two sushi plates. One for Saya and the other for Papa and me, which meant that I'm getting seven and Papa one.

The sushi were small and cut extra thin—tuna, salmon, squid, shrimp, octopus, scallop, flounder, and eggroll with vinegary rice. Except for the eggroll, everything smelled and tasted of a freezer. What's more, the Japanese sushi-men were arrogant and sneered at us. The restaurant was full of beer-drinking Japanese businessmen, some at tables, but mostly at the long counter, where they could order much better-looking sushi. Those top quality sushi had no quoted price. Saya said they could easily cost ten times what we were paying, but those beer-drinking Japanese were businessmen having dinner on company expense accounts.

Late one Friday afternoon Papa brings home a three-foot-long package. He puts it down on the porch and carefully unwraps it. Out comes a fish, a big tuna, all frozen, weighing at least forty pounds, with a large head and a gaping mouth, its eyes blankly staring at us.

Papa works hard to dissect the tuna. Since it's still frozen, he has to use a saw. He cuts the fish into slices, which he puts into our freezer except for two for dinner. How to get rid of the enormous head? Our garbage bin is already full. So we take the head to the garbage bin of our neighbor, the trumpet player at the Cologne Symphony Orchestra, who is on vacation.

Saya is wondering where the tuna came from. Papa proudly tells us he had placed a special order with a fish store in town. The store owner had never seen a whole tuna, because Germans don't eat tuna, except from cans.

Saya thinks this was a great idea. As soon as the frozen tuna has thawed, she cuts it up to prepare sashimi, which she then serves with wasabi, soy sauce, and wakame salad for dinner. In a festive mood, we begin our banquet.

Strange, but the tuna has no taste. It barely smells like fish. Saya sauté-fried the next pieces. No flavor. She grills another one. No flavor.

"Is it really a tuna?" she asks.

"Was it really a fish?" I wonder.

"Yes, it's an authentic tuna," Papa emphasizes, still full of self-confidence. To reassure ourselves, we go out to the trumpet player's garbage bin, all three of us. We slowly open the lid. There it is, on the top of the garbage, this big head, the gaping mouth, large enough to bite off a hand, and blank eyes staring at us. Definitely a tuna.

"The fish store owner assured me it's a wild-caught tuna, brought in from Paris," Papa says, defensively.

"But the fish store owner has never seen tuna before," I remind him, quoting Papa's own story from a day before.

"There are many different kinds of tuna," Saya adds to the discussion. "Some are good for sashimi; others are good only for cans. You have to be an expert. Your fish store owner got the wrong tuna from Paris and probably overpaid."

After this conversation, I never saw the tuna again on our dinner table.

ASCENT OF MAN

Saya appears quite often on nationwide television or we hear that one of her documentary films is running. To see her on the television, or to watch her programs, Papa asked our neighbor, who has a large TV, if we could come over to watch. Now we have our own TV—nice and crisp black-and-white pictures.

Saya checks the programs daily. If she finds something promising, we sit together in the evening on the sofa and turn the TV on. Saya says that from the theatrical point of view, we should be able to judge any program in the first three minutes. We should be able to say yes or no. We watch the program for three minutes and then give a verdict. It's a game and we don't always agree. Most of the time Saya says "no good" and goes back to her room. I stay with Papa to the end. Afterwards we often realize that Saya was right... we should have stopped watching.

We liked the Marx brothers in *A Night at the Opera, The Kid,* by Charlie Chaplin, and a nature film about elephants in Namibia eating fruits. They could not stop putting them into their mouths with their long trunks, but the fruit was over-ripe and fermenting, full of alcohol. The elephants became unstable on their four legs. They hobbled and staggered, lost their balance, squatted down but continued eating the fruits within the reach of their trunks. The music made the scene look even funnier.

One day we watched the first part of a BBC documentary, *The Ascent of Man,* by Jacob Bronowski. It captured me— at once. The author is a mathematician and historian. He embarks on a journey through human history with emphasis on science. He tells how human society developed across tens of thousands of years, from humble beginnings to the modern world.

I have been reading many library books on archaeology and history, on science and religion. The Bronowski series draws me deeper into the world I love and already know a little bit about. He emphasizes how many different factors contributed to human society in its present form—climate and natural resources, the invention of written language, of agriculture and tools, of different philosophies and religions. Bronowski is at ease in front of the camera, with his round eye glasses and graying hair, speaking in a very personal tone, even lovingly, about complex matters. His intimacy with physics and his understanding of life sciences come through in his almost casual presentation.

I want to know more about Bronowski. I go to the city library and, to my astonishment, there is his book with the same title. *The Ascent of Man*, published in 1973. I check it out, take it home, and start reading it, though I'm under time pressure to prepare for the mid-term tests.

All night I read, sometimes nodding in agreement, at other times with questions racing through my mind. I have now finished the first ten chapters. I tell Saya I have a sore throat, and she lets me stay home to finish the book. My grades in the mid-term are surely not good, but who cares?

Bronowski's approach to Pythagoras, Galen, Galileo, Newton, Mendel, Darwin, Einstein, and other scientists is so personal, it's as if he had met them all, discussed with them the problems they were facing, and struggled with them to find answers.

Bronowski brings those great thinkers back to life. According to him philosophy, religion, art, music, literature, and science are all equal in value as endeavors of the human mind. None is higher. None is lower. I agree. I'm passionate about math and about all science that I have encountered so far, but also about French literature to which Bartenstein introduces us in class, about music and drawing, about Grandpa's views of religion, and about my computer. Everything is captivating. No end to discovery. I want to meet Bronowski. I might become his student since I'm soon going to be fifteen.

A few days later Papa tells me that Jacob Bronowski

died soon after the stressful eighteen months of filming *The Ascent of Man.*

BALZAC

"Did you like the story—and why?" Mr. Bartenstein stands there with his arms crossed like a statue of Balzac. "If you didn't like it, it's perfectly all right, but then tell us why." He is very patient. He walks in front of the blackboard from the right to the left and back, looking at each of us. He encourages us to speak up. His gaze challenges the class.

Last week Mr. Bartenstein gave us the assignment to read one of three short stories, *Un cœur simple,* in Gustave Flaubert's book *Trois Contes.*

Now he asks: "Did you like it? If so, why?"

Silence. As we are supposed to speak only French in his class, whether asking questions or answering, few of my classmates feel comfortable raising their hands. Everyone around me opening the book and turning pages.

Since I spent wonderful years at the Chaperon, in the French-speaking part in Switzerland, I have an advantage. At the Chaperon my roommate, Pascal, told me in French all about saints and angels. I had to compete with French boys, stand up in French against Monique, explain in French to Monsieur Bagnoud why I must use the school telephone to call my parents in Cologne. Nevertheless, I had to take a deep breath before raising my hand.

"*Oui,* Mino?" Mr. Bartenstein nods at me.

"I felt sorry for this maid, Felicité, in Flaubert's story," I start. "She has no parents, and no family, and everybody abuses her; I mean, everybody takes advantage of her. She loves everybody and never receives any love in return. Nevertheless she is not angry." I end with a question that has come to my mind while I was speaking. "Is there really such a person in the world?"

"*Merci beaucoup,*" says Mr. Bartenstein. "The question you have raised calls for further discussion. Anybody else with an opinion?" He looks over the class, waits for a moment and nods to Petra, who has raised her hand.

"I've never seen somebody like Felicité. She is surely made up by the author." Everyone in class seems to agree.

Mr. Bartenstein takes this opportunity to give us a lecture about history. "Flaubert wrote the story in 1877, one hundred years ago," he says. "I remind you what Europe was like in 1877, particularly in France, after the Franco-Prussian War had ended just a few years before, with the miserable defeat of the French. The political elite was infighting; farmers, shop owners, small manufacturers, and other business people were hanging in there; but the majority of the French were pretty destitute, barely able to survive. There were surely many women, single women like Felicité, powerless, at the mercy of their employers, disposable and replaceable. In creating the character of Felicité, Flaubert made her indifferent to the cruelty of the society. By saying that she kept the power of love to herself, Flaubert wanted to send a message—that even under the worst conditions, human dignity wins."

Miss Venus comes to visit us in Cologne on her way to a Montessori school not far from here, in the Eifel. It's her third visit and she is always the same—cheerful and dedicated to music. She is disappointed that I have not kept up playing piano regularly. It all depends on a good teacher, I suppose, but I must also confess that I'm now more attracted to drawing.

I tell Miss Venus what I have learned in my afternoon classes at the Music Conservatory. Composers of the Romantic period like Schubert and Brahms. I was startled when I heard their songs. Couldn't believe it. These were the songs Saya had sung to me every night at my bedside. She sang them in Japanese. I thought they were Japanese songs. I was wrong. They are German songs—Schubert and Brahms. Only one of Saya's lullabies was truly Japanese. It had almost no melody and made me sleepy at the end.

Miss Venus sits down at our Bechstein and plays Schubert and Brahms for me. A nostalgic moment. I miss those songs, and I miss my childhood.

———

Mr. Bartenstein recites a passage from *Les Fleurs du Mal,* by Charles Baudelaire. Baudelaire lost his father early in life and had to stay with his mother and step-father, who sent him away to the South Seas. Upon returning to Paris, Baudelaire lived with a black man and wrote poems that were branded as immoral. He was even taken to court in 1858.

In his poems Baudelaire expresses hopelessness, despair, and self-torment; but Victor Hugo praised his poems as an expression of beauty lurking below the surface of evil. Because of his powerful language, Baudelaire left a long-lasting impact, not only on French writers, but also on those in many other countries.

Mr. Bartenstein's dramatic recitation impresses me, though I don't understand why Baudelaire let his protagonist torture himself physically and mentally all the way to the end. His whole life seems to have been an agony, whipping himself for no apparent reason.

We also have to read *Une Saison en Enfer,* by Arthur Rimbaud, the author who started a revolution in poetry and then, suddenly, at the age of nineteen, stopped writing. He drifted through Europe and Africa until his death at the age of thirty-seven. His influence on twentieth-century literature, according to Mr. Bartenstein, was profound.

I like Voltaire, who lived much earlier, from 1694 to 1778. He did not try to be suggestive or mysterious. He was a man of sharp reasoning and sharp arguments. For him God doesn't intervene in the course of human and natural events. Voltaire didn't believe in miracles and divine revelation. I like Voltaire, because he emphasized doubt as a virtue. I like his argument that fanaticism is the most dangerous of all evils. I like Voltaire because he is funny, witty, and spirited when dealing with politics, religion and church.

After we have gone through the nineteenth century and before talking about early twentieth-century writers such as Roman Rolland, Jean-Paul Sartre, Albert Camus, our class takes off on a two-week bus trip to Southern France. Mr. Bartenstein and two other teachers accompany us. We travel through Avignon, Montpellier, and Carcassonne, a medieval town still surrounded by its original city walls and fortifications.

Our destination is Saint-Gaudens in the High Pyrenees. There we stay with French families; I'm assigned to the Rumau's. René Rumau, their son, is my age. I am supposed to practice French with him and his family as much as possible.

René has never heard of Voltaire, Rimbaud, Victor Hugo, Stendhal, Balzac, Baudelaire, or Flaubert. He thinks they are cyclists, big names in the Tour de France.

René loves Coca-Cola, hot dogs, and hamburgers. While I am there, his mother, a heavy lady, serves opulent Southern French cuisine, quite different from the more refined cuisine that I came to know in Dijon in Burgundy, where I had stayed last year with Papa and Saya for the spring semester. While Papa was teaching at the University of Dijon, I attended the public school. We were often invited to dine with Papa's colleagues. Dinner at their houses or in restaurants in town typically lasted from six o'clock to midnight, an occasion to indulge in long discussions, interrupted by many different dishes.

René's room is decorated with posters of Tour de France cyclists. I stopped riding a bicycle after my big accident in

Cologne, when I tumbled over with my bicycle on a rough road and had to be treated in the hospital. I play soccer at the Kreuzgasse during our sports afternoons, but mostly I hang out with Roland, my friend and classmate, who is more interested in art than in soccer. He and I stand around in the corner of the soccer field and talk about our last visit to the museum.

It's hard to find any subject to talk about with René. When I have time, I go out with my sketchbook, looking for a spot I'd like to draw. During the bus ride, whenever we have a stop along the way, I capture sceneries—a rocky landscape here and there, a few farm houses, and some old village churches.

Now in Saint-Gaudens it rains. I take from my suitcase Saya's first novel, *Brokatrausch*, still in manuscript form, three hundred pages long. Before the trip, Saya had asked me to read it but I couldn't find the time. Now, on a rainy day in the Pyrenees, I start with her *Prelude*. I feel I am being transported back to Japan, standing at night at a wide estuary, knee-deep in water, seeing hundreds of paper lanterns floating past me with the full moon high up in the sky. I'm captivated by the story which unfolds. Japan of the early nineteen hundreds. I skip dinner. I finish reading at midnight but still can't fall asleep. Something is missing at the end of Saya's story. Some form of *Finale* that would provide a closure. I think her story must return to the ocean with the full moon. Otherwise her striking opening of the book, her *Prelude*, has no echo.

Back in Cologne I tell Saya.

"Exactly, Minochan!"

She goes straight up to the second floor, to her room with the large office table. She sets out to rewrite the end. Hours later, when she has finished, she reads it aloud to me.

"That's it."

Next morning Saya sends her manuscript to her publisher. He informs her a fortnight later that the first edition will be for twenty thousand copies, a large number for a first fiction.

THAT CHAMPIONSHIP SEASON

Warren Smith, one of Saya's professors at Penn State, comes to visit us with Mia, his wife. I met him last summer. According to Saya, Warren was very sharp in his History of Theater class. He directed two plays each semester at the Schwab Theater, specializing in Bernard Shaw and Anton Chekov.

We take Warren and Mia to museums in Cologne. We drive with them to Maria Laach, the Abbey on the shore of Lake Laach in the Eifel. Founded in 1093, its basilica is one of best preserved Romanesque structures in Germany. It's also famous for having provided political asylum to Konrad Adenauer, the first German Chancellor after the war. Heinrich Böll, the Nobel laureate, whom I have visited with Saya in his house in Cologne, told us about the political brinkmanship the monks of the Maria Laach Abbey had used towards the Nazi regime in order to retain their immunity in face of the Nazi threat. Later, in the Kreuzgasse, I learned more about this chapter of recent German history when we read Böll's novel *Billiard um Halbzehn*.

Warren and Mia ask me if I would like to spend the summer with them at Penn State. What a question! I instantly said "yes," but Papa and Saya are a bit hesitant. Finally, it's decided that they would take me to the Düsseldorf Airport. Warren and Mia would pick me up at Kennedy Airport in New York.

The Smiths live in the outskirts of Penn State. Warren is tall and slender, Mia short and round. He retired a few years ago, but she still works as a typist. "To be professional," she says, "you must be able to type any text in any language. Test me."

I hand her a letter I'm writing to Saya in German and a page from a French book. Then I watch her fingers flying faster over the typewriter keys than I can follow. Not a single typo. She even gets the accents right in the French text.

Over breakfast Warren tells me that Saya was the best grad student he ever had. Her term papers were well structured and pointedly written. He circulated them among the other students. He also tells me that Saya was not good at acting, though she tried hard. She was a good dancer. Warren talks about her month-long nightly performances at Schwab.

"Don't remember all the details, but there was one Japanese dance about a sea turtle...quite unforgettable."

"Urashima!" I cry out.

Warren looks at me with surprise and then continues to tell me that he has been pondering over the meaning of the story.

I tell him it's about Mother Turtle coming back to the same beach every year, but bad boys are harassing her. Urashima comes to her rescue. Out of gratitude Mother Turtle carries Urashima on her back out into the ocean, deep down to the palace in the coral reef, where colorful fish attend her.

"Yes," says Warren, "but isn't there more to that story? For instance, when Urashima is brought back to the same shore, carrying a box he has received from the sea turtle with the promise never to open it?"

I'm lost. Saya never told me about this continuation of the story, but Warren tells me that it's about the concept of time, a recurring theme in all cultures. Between the day the sea turtle takes the boy to her palace in the coral sea and the day she brings him back to the same shore, a hundred years have passed. The passage of time has not affected the boy. However, when he enters his home village, everybody whom he once knew has long since died. Different people are now living there. Feeling alone and lost, the boy opens the box. In that moment, he turns into a frail old man.

For Warren the Urashima story is part of the universal search for understanding time, why we age and die. He says that Saya was able to capture the essence of this transformation from boy to old man by her skillful use of a Japanese theater mask, of sound and lighting together with a change of kimono within seconds. Behind this story, Warren explains to me, is the age-old notion that, outside our world, time is different.

I'm stunned. I've seen Saya's collection of theater masks and the black lacquer box in which she keeps an old man's mask. I never understood the connection. Perhaps Saya wanted me to know only the first part of Urashima, when everything is cheerful and happy. She kept away from me the second part, when Urashima has turned old and filled with despair.

From the moment my travel to Penn State was decided, I've had a secret mission. I ask Warren to take me to campus and leave me there for a day. I walk to the Schwab Auditorium and look for a good spot to sit. I open my sketchbook, take out my collection of pens and the rectangular ink flask, and begin drawing a wide staircase with three arches and the heavy entry doors. Line after line after line.... I try to render the wide stairs and heavy stone building with just two pens of different width. Surely Saya must have run up and down this staircase hundreds of times—autumn, winter, spring, and summer—probably always in a hurry.

After I'm done, I look for the Dean's office. The door is open and I see Dr. Walters sitting at his desk. For a second, he is startled to see me. "Oh, my goodness! Mino. Come in, say hello..." The tone of his voice, as I remember from last year's visit, is a melody that draws you deeply into his sphere. Dr. Walters looks at me with a big smile, and we chat.

"What has brought you here, Mino?"

"Last summer you took us to the play *That Champion-ship Season*, and I saw it once more with my mother. You asked me what I thought about it but we left too soon. Now I want to tell you."

"Well, have a seat."

I pull a piece of paper from my bag, where I have written down my notes. Boring and depressing...I mean not the play, but the human existence it depicts. How can five grown-up men be attached to one single event in their lives, a championship game, which they won decades ago? All five try to give the impression that they are good buddies, while in reality they hate each other. Such a dreary and desolate life. I think it's a depressing play.

"Well, that's how life is, after all—for most people, at least," Dr. Walters says. "That's why we have theater. That's why we create drama, music, and art. To entertain, educate, and enlighten. Have you had lunch? Let's go to the Hub. You've gotten still taller, Mino!"

ROOT OF ANTI-SEMITISM

At the Kreuzgasse Mr. Theil, our history teacher, asks us to write an essay on anti-Semitism. He spent the entire last semester dealing with the Nazi regime, how Hitler came to power, and how he held on to power until the bitter end. One of Hitler's plans was the extermination of the Jews, and he went ahead not only in Germany and Austria but also in the countries under German control during World War II: Poland, Czechoslovakia, Hungary, France, Holland, Belgium, Denmark, Norway, Yugoslavia, Greece, Bulgaria, Romania, and the parts of the Soviet Union that, historically, had been Estonia, Latvia, Lithuania, Belarus, and the Ukraine.

All these countries are Christian countries and have been Christian for a thousand years or longer. In their midst lived

Jews, hundreds of thousands of them. Regardless of whether
the dominant persuasion was Orthodox, Roman Catholic, or
Protestant of all denominations, in every Christian country
anti-Semitism was nothing out of the ordinary. Anti-Semitism
has always been there. Occasionally it seemed to have disappeared in some parts of Europe but then exploded with
renewed vehemence.

Anti-Semitism is by no means Hitler's invention, though
he made full use of it. Why did Hitler want to eradicate the
Jewish people, and why did the people in Europe cooperate
with him, some eagerly, some reluctantly?

When I met Warren Smith at Penn State and joined him
and Mia in a gathering of Quakers, I started to understand
more. Warren calls the Quakers a "Society of Friends." In a
forest clearing we all stood in a circle, our eyes closed, holding each other's hands. There was no pastor or preacher to
give a sermon. The Quakers believe in the inner light that
dwells in all of us. They are against violence—any violence.
They refuse to join any army and any war. They are among
those conscientious objectors who sign up for civil duties
instead.

After the Quaker gathering in the forest clearing, I asked
Warren why Christians and Jews didn't get along in Europe.
They look alike. Shylock in Shakespeare's *The Merchant of
Venice* is considered to represent a typical Jew, and is often
portrayed with a big nose, bent back, and sneaky eyes, but
the violinist Yehudi Menuhin doesn't look "Jewish" at all.
Enough Christian Germans and other Europeans have big
noses, bent backs, and sneaky eyes. So what's the fuss all
about?

For my Kreuzgasse history class essay I write what I have
learned from Warren, a Quaker and professor of history of
Western theater. According to him, there is a long history
of anti-Semitism. For Christians, Jesus is the Son of God,

the Messiah, the Savior, who was sent to this world by the Almighty God to save humankind from all sins. Jews never acknowledged Jesus to be the Messiah. For them Jesus was one of many miracle workers and exorcists of his time, in the first century A.D., calling for repentance throughout Galilee. The end of the world was supposed to be near. Apocalyptic fervor was rampant and the Israelites longed for a Messiah, who could chase the Romans out of Palestine.

According to Christians, Jesus was put to death on the cross by the Jews. Jews, however, maintained that Jesus was crucified by the Roman authorities according to Roman law. He was crucified for sedition and subversion. After all, Palestine was occupied by the Romans and stood under military rule.

Christians labeled Jews "Christ killers." From the Jewish perspective, the Christians confuse their Jesus with the historical Jesus of first-century Palestine.

I learned from Warren that the "Jewish Question" was first addressed in one of the Lateran councils in the Middle Ages. At that time, it was decided that Jews were allowed to live in Christian society, but segregated in ghettos. Jews had to wear certain clothes and were excluded from nearly all professions.

One of the most virulent anti-Semitic pamphlets was written by Martin Luther in 1543. He had tried to convert Jews to Christianity. He wanted them to believe in Jesus as the Messiah, as the Savior, as the Son of God. He wanted to save them from torture in hell as he saw it. But the Jews did not convert. Then Luther became a determined anti-Semite.

I went to the city library and borrowed books on Martin Luther. In one of them, *On the Jews and Their Lies*, Luther calls the Jews "poisonous envenomed worms that should be expelled for all time." Luther says Jews "are no people of God, though they invoke the Old Testament all the time."

He made sure that the Jews were expelled from Saxony and from many German towns.

I couldn't believe what I was reading. I had a favorable view of Luther, who was the first to translate the New Testament into German, who started the Reformation in Europe, who propagated a theology of love, who criticized the Papal practice of selling indulgences. Yet such a great person was able to hate Jews to that extent only because they had a different view of Jesus.

Where does this intolerance come from? To believe in one thing, why is it necessary to reject everything else?

The more I read, the more I learn that this intolerance has been endemic across the Christian world, not only in Western Europe, but also in Eastern Europe and in Russia under the Orthodox Church, and in Poland under the watchful eye of the Roman Catholic church. Many pogroms were conducted, mass slaughters of Jews, repeatedly and fervently carried out by Christian mobs.

Warren Smith introduced me to the fact that during the second half of the nineteenth century and well into the twentieth century, Darwin's theory of evolution had been applied to human society and human races. Known as "Social Darwinism," it morphed into race science. Its representatives declared Jews to be biologically inferior to the Aryan race. This provided legitimacy to the Christian feeling of superiority. The whole of Europe and America embraced race science, but the movement was strongest in Germany and Austria, where an Aryan paragraph was written into law in 1885. Using the Aryan paragraph, Jews were excluded from all kinds of associations such as sports clubs, choirs, and fraternities. Anti-Semitism became a way of life in Europe—but not only in Europe. Also in America.

Hitler used the Aryan paragraph to declare not only Jews but also people of the Slavic countries to be racially inferior.

The theory of race science extended to all people of color in Asia, Africa, and non-white America.

One problem that Hitler faced was his alliance with Japan, a country inhabited by an authentically yellow race, and therefore considered definitely inferior to the Aryan race. To obscure this dilemma, Hitler diverted the people's attention to their own lineage. Only those of Aryan descent without Jewish parents or grandparents could enter government service, be employed in schools, universities, theaters, and orchestras, or be employed in banking and in the press. In short, Jews were driven essentially out of all professions.

Germans had to prove with Aryan certificates that they didn't descend from either Jews or Slavs. They had to prove that their parents and grandparents were all decent, pure Aryans. They had to show their birth or baptism certificates.

Where to find birth or baptism certificates from two or even three generations ago? In the church records. The cooperation of the Church was essential, especially in the countries occupied by the Germans during the war. With ready cooperation, primarily from the Church, the German SS was able to round up the Jews within days of German tanks rolling in.

In the meticulously tended city park not far from our house in Cologne stands a small stone monument partly covered by bushes. It reads: "Here the Jewish Men, Women and Children of our Community were Assembled for Transport to the Death Camps."

What did Hitler gain from this?

What did the German people gain from this?

———

Mr. Theil, our history teacher, didn't like my essay. He gave me a bad grade. He told me that Hitler's anti-Semitism had nothing to do with the Church, because Hitler was not a religious man.

I told Mr. Theil that he was mixing up cause and effect.

The centuries-old prejudice of Christians against Jews was the cause of the nineteenth-century pseudo-science of human races. Without more than a thousand years of hatred, deeply rooted in Christianity, Hitler could never have achieved his sinister goal.

Mr. Theil didn't change his mind.

A LIGHT GOES OUT

Saya has flown to New York, because her novel *Brokatrausch* has been translated into eight languages, including English, *Samurai*. Joan Davis, her literary agent, has arranged several events for her in New York, Philadelphia, and Washington, D.C.

I sit with Papa at home discussing possible universities I should visit before we make a final decision about where I should apply for the fall. We think of Oxford and of Cambridge, where Dr. Needham has already mentioned the names of some mathematicians and physicists for me to meet. We think of nuclear physics at Orsay in Saclay and of the École Polytechnique in Paris, as my French baccalaureate from the Kreuzgasse allows me to enter any French university. We also talk about the Swiss Federal Institute of Technology, ETH, in Zürich, the MIT of Europe, Einstein's alma mater. It's reputedly the most difficult place to get into and to make it through the first two years.

Saya calls to tell us everything went well with her events in the USA. She is now on the way to London to meet her British publisher. Relieved, Papa and I go to bed.

In the morning Saya calls from her London hotel, out of breath. "Wrong luggage...huge sandals and a man's swimming pants...three packs of baby diapers, baby powder, and baby cream from Johnson & Johnson, a big can of powdered milk." We can hear her digging through the suitcase that is not hers. "Looks exactly like my bag, beige with green and red stripes…four wheels at the bottom. I bought it in Kyoto just three months ago."

"Does the bag have a nametag?" Papa asks.

"No…no, oh yes…yes, Bhatta…Bhattacha…Bhattacha-raya, clearly an Indian name, but no address."

The following hours we spent calling back and forth between Cologne and London. Saya is upset with the British Airline customer service, which replied haughtily with no willingness to help. A representative of the Lost and Found noted mechanically Saya's name and the name of the hotel and its telephone number. Now Saya is really agitated. Two of her best dresses which she had packed for the trip. A pair of elegant high heels and an expensive pearl necklace. What if she didn't get back her notebook, where she has scribbled down the names, addresses, and phone numbers of every person with whom she has interacted over the past thirty years?

I've never experienced Saya in such a panic, perhaps because I couldn't see her but just heard her choking up. In the past, if I had a problem, she fixed it. Anytime. Anywhere. Now she has a problem, and she sounds so helpless.

Papa says, "Don't worry. You are safe. No plane accident, and we are both here."

I also want to comfort her: "*Daijobu yo, Saya!*"

Late in the evening Saya calls from Piccadilly Circus, where she has just met and exchanged identical bags with a young Indian, who was also very relieved and happy to get his luggage back.

———

I take the train to Paris to visit Mr. and Mrs. Gasser, whom we had met in Crans-sur-Sierre while I was still at the Chaperon. The Gasser's live not far from Versailles. I go out every day with my sketchbook and a lunch bag, to draw landscapes and cityscapes that catch my eye.

Versailles looks bombastic. Having been registered as a

World Heritage site by UNESCO, it is overflowing with tourists. I follow a guide who shows us room after room in this elegant Baroque architecture, Gobelin tapestries, Rococo furniture and chandeliers. I'm startled to hear that the original chateau had not a single bathroom or toilet. Kings, queens and all the noble people of the time never took a bath, but instead applied all kinds of perfumes.

Notre Dame de Paris, the Gothic Cathedral on the Île de la Cité, is fascinating. The most impressive church I've seen so far, a beautiful testimony to the medieval trust in God. I take a seat at the middle aisle, close my eyes and listen to the sounds. It's as if I could hear the voices of angels. I pull out my block and start drawing a side altar with two kneeling figures, their backs bent, showing how lonely and helpless they feel.

Saya and Papa also come to Paris and we visit the École Polytechnique in Paris and then Orsay in Saclay, two universities that I am considering for the fall.

Next day Saya and I spend a full day with the Impressionists in the Musée d'Orsay. We walk from room to room just to get the feeling of the display and then return to the rooms of our favorite paintings and drawings. We stand in long silence, absorbing the colors, lines, compositions. It's something I can hardly express in words. Saya asks my opinion as to the perspective of certain paintings.

After Mr. Bartenstein's French literature class and after reading works of the nineeenth-century French writers like Victor Hugo, Flaubert, Balzac, Baudelaire, and Rimbaud, I imagine how the Paris society looked when Monet, Renoir, Cézanne, Courbet, Manet, Van Gogh, Gauguin, and many others lived and painted here. Recently I've begun experimenting with watercolors. How difficult it is to create light and shadow with colors. The paintings of the Impressionists dazzle me. Walking down the street, I start looking for them.

Nine years at the Kreuzgasse are nearing their end. Over the course of the past nine years, more than one third of the students have left for other schools, mostly vocational schools that emphasize more hands-on education in different professions. The ones who persevere are university-bound.

Everybody in class is nervous. The finals will be taken in rapid succession—math, physics, chemistry, geography, biology, German, French, English, and history. Whether or not we receive our abitur and baccalaureate certificates depends on the outcome of those exams.

Many of my classmates cover their anxieties by smoking, fiercely blowing smoke during intermissions. Some are excessively loud, romping around for no reason. The rest are going over textbooks and notebooks with pale, stale faces.

Out of the blue, in the middle of our nervousness, the news breaks: Mr. Bartenstein has collapsed. He has been transported to the university clinic and from there to the psychiatric ward.

No, that can't be true. I must go to see him at once.

Someone shouts: "Hey, who is going to give us the French exams?"

Others shout back, "French is out!" Many sniggers. "Hurray, Bartenstein is out!" The whole class erupts in laughter.

I go to the director's office and ask the secretary where I can find Mr. Bartenstein. She barely looks up from her typewriter: "No visitors allowed."

———

Coming fresh out of graduation, the first thing I do is to walk over to the University Psychiatric ward. It's a hostile-looking building, heavily fortified. The uniformed guard at the entrance informs me that nobody is allowed to go in, except for family members or with a written permission of the doctor in charge. I ask for the name of the doctor in charge, but the guard won't give it to me.

At home I ask Papa for help. As a university professor, he can talk to the head of the psychiatric ward. It takes a few days, but at the end I receive the permission to enter the building. Papa comes with me. We are led through three metal doors, opened and locked again. We enter a long corridor and then a small room with a semi-dark ceiling lamp. There, along one side, we see a white hospital bed and an almost lifeless body.

I call "Mr. Bartenstein!" and take his hand. He looks at me with a blank stare and says nothing. I tell him how much I have missed him in school and that I don't feel joy after having made it through the abitur and baccalaureate. I tell him I wanted to celebrate with him, because he is the one who encouraged and inspired me at all times.

Mr. Bartenstein continues to stare at me but I detect in his gaze a slight sign of recognition. I want to take him out of the room into the park, because it is depressing to be caged in such a gloomy place. Papa goes out and talks with a nurse. After some phone calls, she gives us the permission to be out for one hour.

We put Mr. Bartenstein in a wheelchair and go out into the Green Belt park. It's a warm summer afternoon. Papa pushes the wheelchair and I walk next to Mr. Bartenstein, talking to him. He has become small and fragile and suddenly old. I wonder whether, now, in the park, he notices the change in his surrounding. His face remains frozen and he doesn't utter a word. So, I keep talking without expecting a reply, telling him that I've sent out my graduation certificate to the Department of Physics at the Federal Institute of Technology in Zürich. ETH is reputed to be the most difficult place to get in. It's known to have a tough entrance examination. With my abitur and my baccalaureate, I could enter any university of my choice in Germany and France—without any entrance examination. Why ETH? Because it is the best. I continue to make an effort to talk cheerfully to Mr. Bartenstein to suppress my sadness, hoping that he can understand me a little bit.

I ask him if there is anything he'd like me to bring to his hospital room. I ask whether the food of the hospital is all right, and if he is warm during the night. He looks at me without uttering a word.

Papa glances at his watch and gives me a sign that we must return.

I tell Mr. Bartenstein that I'll come back to see him again and, if the weather is fine, we'll go out into the park. I'll bring a blanket for the night and a bag of French cookies to eat together. At the entrance to the psychiatric ward the nurse is already waiting for us. She resolutely pushes his wheelchair through the first of the locked doors.

What a great teacher he was, full of love for the French language, who taught me what literature is, what it can be, and how the literary mind works. He was always ready to listen whenever I had questions or comments. His intense look encouraged me to think further. Now he has been reduced to a soundless bundle. I'm so sad. I'm full of sorrow.

On our way home in the car, Papa says the department head didn't give him an explicit medical diagnosis, but hinted Mr. Bartenstein must have been suffering from war memories. He was one of the youngest to be drafted into the army in 1945. The worst atrocities happened at the end. Bartenstein must have witnessed unspeakable things and buried their memories deep in himself. Then, suddenly, after all those years, something must have gone awry in his brain.

For the next three months, I visit Mr. Bartenstein every afternoon, rain or shine. Every morning, I work at home for the ETH entrance examination, math, physics, chemistry.

After lunch with Saya, I go by streetcar to the university psychiatric ward. The nurse has become friendly to me. I talk to Mr. Bartenstein casually as if nothing has happened. He looks at me, though he utters no word. If the weather permits, I push his wheelchair through the park. I ask him trivial questions and answer them myself just to fill the emptiness, which otherwise would turn into pain.

Mr. Bartenstein, you encouraged me from the beginning. You were the only one of my teachers who ever said how wonderful it is that I have a Japanese mother, how great that I grew up in two, even three cultures, that I can be proud no matter what some others say.

Once Saya comes with me. While I push the wheelchair, she tells Mr. Bartenstein about her next novel. She reminds him that he wrote brilliant reviews of her first two novels, the most professional and balanced she had ever received.

Saya steps in front of the wheelchair, squats down and puts her hands on Mr. Bartenstein's shoulders: "Please come back. Please come back." She starts to weep. He looks at her face. Then a sad, endlessly sad smile appears.

On one rainy afternoon, two days before I have to leave for Zürich for the ETH entrance examination, I sit at his bedside in the semi-dark hospital room. I tell him about the concert I've heard the previous night. Suddenly, Mr. Bartenstein sits up in his bed, points at the door and screams at the top of his lungs:

"Stop! Stop! Don't shoot! They are children...Children, they are children! Don't shoot! They are children! No! No! No!"

His voice breaks into a whimper. Before I can catch him, he collapses to the floor. I take him in my arms. "They shot them all...shot them all."

The nurse comes. She gives him an injection. His convulsions subside. I help her lift Mr. Bartenstein up from the floor and put him back in bed. The nurse nods at me approvingly and then asks me to leave.

On September sixth, Saya gets up early in the morning, makes my lunch-box with *onigiri* and takes me to the main Cologne train station for Zürich. When the Express rolls in, she comes with me into the compartment, takes me tight in her arms: "*Daijobu yo*, Minochan!" She swiftly steps back onto the platform. There she stands, waving until the ever-winding train leaves her at the last curve.

THE TOKAIDO ROAD

The day after I returned from Zürich, I leave with Saya for Japan. She purchased the least expensive tickets long ago, packed two suitcases with many presents, my sketchbooks, pens, watercolors, brushes; and our clothes for three and a half weeks. It's early autumn, the best season in Japan, except for the typhoons that are expected every year—as many as twenty or even forty. Saya says typhoons strike only parts of Japan, for instance Kyushu, the Osaka-Kyoto area, and the Yokohama-Tokyo district. Some typhoons come with immense downpours, others with frenzied storms, many with both. Despite their violence, the typhoons calm down after half a day or so.

Papa, who has just returned from a conference in the USA, drives us to the Brussels Airport. We are going to fly Sabena Airlines over the North Pole to Anchorage in Alaska and be on the ground for an hour for refueling. I hope I'll see polar bears and Inuits. Our flight is fully booked but I have a window seat. In the back of the plane the passengers are smoking. The bad smell reaches us nonetheless. Saya says it's still better than Lufthansa, where the smokers sit on the left side of the plane and non-smokers on the right side.

I think of Mr. Bartenstein. How good that he is recovering. I heard that he can talk again. Before leaving I learned that a social worker is coming to the psychiatric ward to prepare him for release. Saya says he'll surely be much better by the time we come back.

The flight attendant brings our dinner. It is not as good as we expected. "The Belgian kitchen is reputedly even better than the French cuisine," Saya says. We unpack our Japanese lunch from home and start eating it. "The worst is Aeroflot," she adds, biting into her *onigiri*. "There you get a large chunk of meat, so tough that you can chew on it all the way from Moscow to Tokyo." The advantage of Aeroflot, however, is that the distance from Frankfurt to Tokyo over Moscow is the shortest, only twelve hours, instead of sixteen to seventeen hours via Alaska. And Aeroflot is cheaper. That's why it's booked out months ahead of time.

We arrive at Anchorage and must leave the plane to go into a large waiting room in the terminal. In the center stands a huge stuffed polar bear. Everywhere handicraft of Inuits; brand-named ties and scarves from Italy; winter jackets and sports goods; jewelry and cosmetics. We are not allowed to go out of the airport building. I can't see any polar bears or Inuits. As I am about to fall asleep out of boredom, we are called to get back into plane.

Saya asks me why I want to study physics. Her sudden question wakes me up from my slumber. I tell her if I make it into the ETH, the first two years will be entirely math.

"But you've done enough math already. Why do you need more?"

I explain to her in a way she can follow. I tell her that the essence of physics is to find relations between natural phenomena from the tiniest atoms to the vast distances of the universe. In order to describe these relations, physicists use mathematics. That's their language. That's why physicists can communicate with each other across all human languages and cultures—using math. Math is universal.

"Like music?" Saya asks with a happy smile.

"Yes, yes, but music is emotion. Math is logic."

"But you have already done so much math in the Kreuzgasse."

"That's just the beginning…a tiny, tiny portion of an infinite space."

"How old is math? Hundred or two hundred years? That old?"

The question surprises me. I tell Saya, "At least five thousand years old in Mesopotamia and India. At least thirty-five-hundred in China. The fact that we count twenty-four hours to a day, sixty minutes to an hour, and sixty seconds to a minute, comes from Mesopotamia, from the time of the Sumerian civilization in what is now Iraq. The concept of zero and the concept of infinity come from India."

"What is cosmology?"

I tell her that cosmology is the study of the universe, how it may have started and expanded over time. I tell her how stars are born, how they age and die. "To explain all this," I tell her, "We need math. With math I can write down the law of mutual attraction. Mathematically it's the same between atoms and between stars and even galaxies, though the scales are vastly different."

I look at Saya to see if she is following my explanation, but she is sound asleep.

———

In Tokyo we stay in the Shiba Park Hotel. Saya brings me to Akihabara, where there are hundreds of stores for electronic goods. She is bored by all the blinking lights and gadgets. Then she takes me to the Meiji shrine and its expansive garden, where one hundred seventy thousand trees from all over Japan were planted some fifty years ago. A forest in the middle of the megacity.

We walk to Asakusa, the amusement district, past the Sensoji Buddhist temple with its gatekeepers, the Thunder god to the left, the Wind god to the right. Both try to look furious and threatening, but I find them humorous. The original Sensoji temple was built in 1648, destroyed by the fire bombing of 1945, and rebuilt after the war. A tiny Bodhisattva, barely four centimeters tall, is enshrined in its inner sanctum. According to legend, three fishermen caught the figurine in their nets in the Sumida River, long, long ago.

Many people, mostly province folks and foreigners, are strolling through Asakusa. Countless souvenir shops, bakeries, bonsai displays and eateries are lined up and magicians display their tricks to draw customers. In the middle of seventeenth-century Asakusa was already a bustling amusement district in Edo, then a city of two-million—four times the size of London at that time. Asakusa kept growing until 1923, when the big earthquake destroyed it, together with most of Tokyo. What was then rebuilt became ruins again, this time by bombings in 1945.

We go into the Kabuki Theater to watch the afternoon program. Inside a large hall, several stories high, with row of galleries. As a student Saya came here to the top gallery, standing room only. In Kabuki theater all roles are played by male actors dressed in multi-layered, colorful kimonos and wearing heavy makeup and wigs. They talk in singsong. Their movements are choreographed and their stories are familiar to the public, just like the stories of Western operas. Hard to believe that the dainty young girl I see gracefully dancing on stage is actually played by a seventy-year-old male actor.

———

We travel to Kyoto by bullet train, eating lunch from boxes, rectangular and oval, we bought at one of the opulent food shops in the Tokyo station. Saya tells me that, during her student years, the trip took eight hours and the trains ran just a few times per day. Now the trip takes only three hours, and the bullet trains run every ten minutes. Back then, Saya journeyed from Kyoto to Tokyo four times a year for four years. She got out of the train at any of many stations along the way to buy a lunch box. After eight hours in a train, her face was blackened and her clothes smoke-stained because the locomotives still ran on coal.

In the summer of 1958, when she came to America, Saya took the trans-continental train from Oakland, California to Pittsburgh. Outside the temperature was burning hot with the sun blindingly bright, but inside she was chilled to the bone. For the first time in her life, she experienced air condition- ing. Three days and nights. All other passengers seemed to know what to expect. They had jackets and carried heavy blankets. They carried baskets full of food. Saya had none because she thought there would be plenty of food to buy on the train—like in Japan.

Somewhere in Nevada a middle-aged woman came into the train and sat next to Saya. She covered herself with a big blanket and took a sandwich out of her bag. She noticed that Saya was freezing and put her hat on Saya's head. She shared half of her sandwich with a smile and told the other pas- sengers around that they should also feed this Oriental girl.

At that time a Japanese was still a rarity in America. One after another, the passengers came by to inspect Saya, who was sitting cross-legged on her seat, covering her bare feet under her summer dress, and looking around from under- neath the wide-brimmed straw hat. One by one the passengers gave Saya a slice of home-baked pie, half a sandwich, biscuits, chocolate, and a cup of hot chicken soup from a Thermos jug. Then came a piece of cheesecake, more slices of pie, and two hot dogs. Saya swallowed them all in order not to disap- point the onlookers. "Oh, isn't that something." "She enjoys my cake." "Look, look, she has eaten it all." Saya felt like a monkey in the zoo, but kept smiling until she had to rush to a bathroom, where she threw up all she had gulped down.

When the woman next to her got off at Salt Lake City, she took her hat back, but left the blanket with Saya.

The bullet train is passing through the Hamamatsu Station. Mount Fuji should be to the right but clouds block the view. I think of Hokusai's woodblock prints of the Fuji silhouette from daring angles and bold perspectives. Hokusai published his prints almost one hundred fifty years ago. Yet they look strikingly modern. The villages and towns were different. Japan's population was thirty million then instead of one hundred twenty million now. The Tokaido was the main road connecting Edo, today's Tokyo, with Kyoto and Osaka. It was a heavily traveled thoroughfare, used by more than two million people every year, with fifty-three way stations, which served as relay stations for the postal service. The way stations were clusters of inns and eateries. There travelers on foot could rest or sleep, and horsemen could feed or change their horses. In travel journals written by Portuguese, Dutch, and German visitors in the late 16th and early 17th century, we can learn how well kept the broad road of the Takaido was, how crowded it was any time of the year with all kinds of people—from daimyos and samurais down to merchants and peasants, and how spotlessly clean the wayward inns. The Europeans were particularly impressed by the fact that they had seen many women traveling in groups along the Tokaido, obviously safe from violence.

Listening to Saya I think of Europe during the late sixteenth and early seventeenth centuries. Portugal and Spain were expanding their colonies across Central and South America, across the Philippines. England and the Netherlands established their East Indies Companies. England and France were racing to grab territories across North America. France was fighting England over control of India. On the European continent, religious wars between Protestants and Catholics were commonplace, the worst being the Thirty Years' War, which involved Austria and Spain on one side, Prussia and Sweden on the other side, with France coming in and out at will. According to one source, eighty percent of the German population perished during the Thirty Years' War.

Surely no tourists could have traveled across central Europe during that era, enjoying food, bath, and safety as the Japanese did in their own homeland around the same time.

However, looking back and disregarding the human suffering, I accept that it was also a glorious time. I think of Shakespeare who, in the same century, wrote play after play that continue to enlighten even today's audience. I think of Claudio Monteverdi, a composer of sacred and secular music, who wrote the first opera, *Orfeo*, in 1607. I think of Galileo, who was convicted of heresy by the Inquisition in 1632 for having defended the heliocentric view. I think of Rembrandt van Rijn, my favorite painter, who created magnificent character portrayals such as *The Night Watch* in 1642. I admire all his works, particularly his self-portraits as a young, middle-aged, and old man, which reveal so much of his inner self.

I tell Saya all what comes to my mind. Europe is so full of contradictions, full of cruelty and beauty, complex and bizarre.

We arrive in Kyoto.

I spend the first day with Grandpa in his shrine on the hill. Next day I go to the Daitokuji temple, my sketchbook in hand. There are many buildings within its vast precinct. Seven years ago, Saya took me to the stone garden. Meticulously raked gravel. No trees or flowers, only moss covering a few boulders. I saw Zen monks cooking rice in a huge iron cauldron with firewood.

Today I walk under pine trees. Some young monks in loose black outfits are sweeping the path and a group of Japanese tourists walks by, hurriedly bowing to them. Saya told me that every morning before dawn, the Zen monks sit together to meditate for an hour. After breakfast they meditate again, read sutras, and sweep the trails. I'm not sure I want to be one of them.

I'm drawn to a closed wooden entrance door. I sketch the long narrow stone path leading to the door and curved roof rising behind the trees. The branches are slightly shaped so that they follow the curvature of the roof. Elegance without weightiness.

I sketch some other spots where nature and temple framework support each other to create stillness—a delicate, yet sophisticated interplay. The Daitokuji temple was built in the middle of the fourteenthth century as one of the great Zen temples in Kyoto. About the same time, the gothic sanctuary of the Cologne Cathedral was completed with its high arches. What a difference in approach. Both are places where people try to meet some higher beings. Personal or impersonal. Through meditation or prayer, through ceremonies, rituals, songs, music, or whatever.

Saya has told me that in Japan, people don't have to commit to any temple or shrine. They don't have to belong to any one organized religion or faith. They can belong to two or three religions at the same time. Nobody frowns if they belong to none. They are free—and have been free throughout history—to visit any temple or shrine and to pray to any of the deities residing there.

What a difference to Europe. In Christian Europe, for centuries past, everybody—each man, woman, and child— had to belong to one denomination, to one denomination only, be it Catholic or Protestant. Each church controlled its flock of people. Each church was intent on keeping its congregation on a tight leash. Changing one's religious affiliation was a serious matter, a conversion deemed irreversible.

I go into one of the Daitokuji buildings surrounded by a

thick bamboo grove. A few visitors sit on the veranda listening to the wind rustling the bamboo leaves. I wish there were pandas climbing the bamboo stems.

The Nijo castle fascinates me. It's set in a large garden, surrounded by wide moats. Built in the early seventeenth century, it was the palace where the Shogun stayed when he came from Edo to Kyoto to confer with the Emperor. I sketch a few views of the buildings. I take a tour. There are bathrooms and toilets in the Nijo Castle. I think of Versailles, built around the same time, and smile.

THE LAST TIME

My sketchbook is filled with scenes of fishing boats anchored at a pier, of waves breaking on rocks, of a little island not far from shore, and of towering clouds on the horizon. Ten days in a fishing village on the Pacific coast. I can't use the easel Saya borrowed from her friend in Kyoto, because sudden gusts of wind coming from nowhere toppled it together with a canvas I was working on. I gave up watercolor and stuck to drawing.

At the village inn overlooking the ocean we had gorgeous dinners every night, at least six courses served in our room. Breakfast was also opulent, every day offering different varieties. In the middle of eating abalone and lobster tail Saya lets out her worries about what Papa might be eating back home in Cologne. Before leaving she prepared food for him and refrigerated it, but it's surely gone by now, she is thinking out loud. I tell her Papa is happy with a piece of bread, with a bit of butter or cheese, a few stalks of parsley, and a handful of nuts; that's enough for him. Saya says that's not a meal; he is surely feeling miserable.

I'm stunned how little she knows about Papa's Spartan eating habits. Not only him, but most Germans. They are content as long as their stomachs are filled—bread, butter, and sausage. When they eat hot meals, they put gravy over everything—meat, potatoes, and vegetables. During my school trip to southern France, my classmates were unhappy with the food served by our French host families—delicious meals with a hint of herbs and spices. My classmates complained about the horrible French cuisine. No gravy on the table.

––––––––––

Back in Kyoto, I spend days drawing Kenkun Jinja, Grandpa's Shinto shrine. I start with the structure that is open on all four sides. Maple branches with lustrous green leaves touch the roof. Bright morning light falls on the wooden platform with its smooth non-glossy finish. I marvel at the craftsmanship. The wood is breathing as it reflects the light. It's not easy to appreciate plain wood and its luster absent of any paint or varnish. Plain wood may not look precious, but only the best quality lumber can hold a non-glossy sheen and withstand the test of time.

I am reminded of Saya's childhood story of that September day soon after Japan's capitulation that ended World War II. It was her first encounter with Americans. She was ten at that time. She was polishing the wooden floor of this very same open structure with rice bran. She heard loud male voices approaching. Three tall men in beige-khaki colored uniform appeared at the bottom of the stairs. Never had she seen such giants before. With a few big steps they were up on the stage—with their shoes on. They started walking across the wooden floor she had been polishing for the past hour.

Angry and upset, little Saya stood up. "This is a sacred place," she said in Japanese. "You are not allowed to step on this floor with your shoes." She spoke resolutely but calmly, as her father would have done. The three giants looked at her. She stared back. Somehow, they understood. They smiled coyly and retreated.

After lunch with Grandpa I continue drawing. The main shrine, with two tall stone lanterns left and right, stands in the sunlight. I focus slightly from the side. The stone lantern on the left moves to the center with its moss-covered column and well-proportioned groundwork. Lightly curved front roof of the altar becomes the background of almost the entire composition.

People come, lost in thoughts, bow before the altar, clap their hands twice, murmur a few words, bow deeply once more, and leave. I continue undisturbed.

Time drifts. The sun sets behind the altar making its silhouette lucid. Grandpa comes and brings a cup of green tea and an incense that repels mosquitos. He lights it and places it by my feet. Then he stands for a moment looking at my drawing. *Ah, sugoi na*, he says as if to himself. He strokes my head gently like Saya always does when she wants to show affection. Then Grandpa goes back as quietly as he came.

Saya has left for Mt. Hiei on the other side of the city. She wants to talk to the abbot of the Enryakuji Temple, Grandpa's good friend. Last time when I was in Kyoto at the age of twelve, she took me to Mt. Hiei by local train and cable car. I remember it was a hot day. We walked between the old temple buildings. Mt. Hiei is famous for its many birds. We heard them chirping and twittering in the trees as if they knew that they have been declared a nature treasure in 1911 and protected ever since.

Now I'm spending the whole evening with Grandpa. After dinner we move to the East Room, where we have a commanding view of the entire city and the chain of mountains beyond. The city reveals itself in purplish-violet shadows and thousands of flickering lights.

"Sixty years ago, to this very day," Grandpa says with an almost casual voice, "my father died. He was only forty."

It sounds as if a page from an ancient book has started to speak. I'm quiet and wait for more.

"I was fourteen at that time. My father and I were out on our daily horseback patrol. We had ridden our horses over narrow paths between rice fields, around mulberry trees, and over the bridge to the gently sloping hills. We had enjoyed the balmy evening air. As usual, we were both riding without saddles, and the horses seemed to like it too. Approaching our house where my younger brothers were playing in the yard, walking on stilts, my father called out 'Dinner time!' I heard my mother echoing from inside the house, 'Dinner time!'

"We reached the stable, and we both got off our horses. I led mine to the water bowl and patted his mane. Just in this moment, my father grasped his chest, cried out in pain and collapsed. That was it. In front of my eyes. He was dead."

Grandpa looks over the city where the lights multiply while the color of dusk fades away. After a while, he simply says, "On that day my quest began."

As the evening darkens, Grandpa tells me about his quest.

"Where did my father go, a tall, slender man so full of vigor?"

Grandpa started to visit Shinto shrines to seek an answer. In vain. He started to go to every Buddhist temple he could find. In vain. When he became a student in Tokyo, he started to go to Christian churches, Catholic and Protestant, listening to their sermons and the recitations of their Holy Scriptures. In vain.

According to Buddhist teachings there is Nirvana, the Blessed Land, where everybody goes after death and lives happily ever after. The problem is that the historical Buddha never mentioned Nirvana, and he never spoke about life after death. His search was directed to life as we experience it, and how we can free ourselves from fear, resentment, and grief. The idea of Nirvana was a Hindu element that became integrated into mainstream Buddhism long after Buddha's death. It was added in order to address people's fear of the end of Life and the uncertainty of what comes next.

Likewise, the Christians have Heaven where all those who have done good during life will go and live forever in the nearness to God and Jesus. The Christian God, the almighty Creator of the world, has evolved out of the Jewish God, Yahweh, the God of Moses, the God of Wrath. Yahweh was the God the historical Jesus knew, but he changed Him into the God of Love, the benevolent Father in Heaven to whom Christians go after death.

In Shinto, life is wholly of this world. Life is the one and only chance we have to be on this planet. Life is a precious time, best spent to do good deeds, for which you can be remembered. Death is inevitable. The end of life. What remains are memories. There is no afterlife in a far-away imaginary place—no Nirvana, no Heaven, no Hell. The souls of those who have passed away live on, if they are remembered and as long as they are remembered.

Shinto replaces eternity by the concept of memory—memory in the minds of those who live after us. As long as they remember us, our souls live on. When the memories fade away, our souls will also fade away, merging back into Nature, from where we all come.

During his search as a young man, Grandpa met a Taoist philosopher who pointed him to the Chinese Book of Changes, the I-Ching, thought to have been composed around the same time as the Upanishad and Rig Veda in India, as the old Sumerian texts in Mesopotamia, as the Old Testament in the Middle East. That is how Grandpa's quest began to decipher the intricately interwoven, multilayered mystery of Life.

"The I-Ching goes back to a time before the first clay tablets were created in Babylon," Grandpa says, "before the first hieroglyphs were written in Egypt, and long before the Greeks appeared on the stage of European history. The I-Ching contains knowledge accumulated in ancient China over many centuries, probably millennia. The I-Ching is a work of mathematics that applies algebraic techniques to combinatorial problems, fused into one monumental system of symbols."

Grandpa notices that I'm suddenly leaning forward to catch every one of his words. Though it is already pretty dark, I see his smile.

Now I begin to grasp why there are so many books in his study room, which Saya showed me, rows after rows, mostly books on the religions of the world—Buddhist sutras of the early period and later times, Sanskrit texts, the Old and the New Testament, various Taoist books and the Nine Chinese Classics, several editions of the I-Ching, the Book of Changes, with thick commentaries.

I am reminded of a conversation I recently had with Saya. As I was sitting in Grandpa's office, drawing the desk, where he normally sits, Saya pulled out two old books, printed on traditional Japanese paper and sewn by hand with twisted silk thread. When she opened the pages, she found many notes in neat small letters and some comments that Grandpa had jotted on the margins. Here and there were big Japanese letters written in a childish handwriting across the pages.

"That was me," Saya said with a self-conscious smile, "when I was four and just learning how to write." She told me that Grandpa never scolded her, even though—as she later realized—she had been scribbling into one of his most precious books, several hundred years old.

Grandpa talks about the I-Ching and that—for him—it was the gateway to unravel the intricately interwoven, multilayered structure of Life. An enormously complex task to understand the combinatorics that the Chinese sages had developed over the centuries—a language of numbers holding information about things that influence and control every fiber of Life.

I heard from Saya that Grandpa had written a series of books about the mathematical structure of the I-Ching and the interpretation of its symbols. These books, three of them, published before 1935, have become, so she said, the corner-stones of the interpretation of the I-Ching. They are being used by scholars throughout Japan and have been reprinted several times.

Grandpa stretches out his hand to the East: "Look."

A full moon is rising over the mountain range beyond the city, exceptionally large and orange. The night sky is gleaming, ink black. It's an almost dazzling sight. In my mind I hear Beethoven's *Moonlight Sonata*.

Grandpa looks at me with his gentle but piercing eyes. Then he says, "Minochan, remember what I told you tonight... because this might be the last time we can be together."

UNIVERSITY

ETH, ETH, ETH

Papa was waiting for us when we arrived at the Brussels Airport. On our way home in the car, I start to tell him what we did and saw in Japan during the last three and a half weeks. Papa interrupts me, saying, "Oh by the way, Mino, the fall semester at the ETH begins in a week."

"Am I accepted?" How often had I been worried in Japan about the outcome of my entrance examination.

"No question." Papa laughs, as though he knew it from the beginning.

"Since when did you know? Why didn't you call us immediately?" Saya's tone is reproachful.

"Oh, I completely forgot about it. Never expected otherwise," Papa continues in a matter-of-fact tone. "By the way, Mino is the only one from Germany who was accepted into Physics. Twelve others applied, so I heard, all top grades from other German schools, but none of them made it."

Saya remained angry for the next few minutes, but then sang out, "Minochan! *Minochan!* ETH, ETH, ETH, hurray, *hurray!*"

I'm overwhelmed and relieved. As soon as we get home to Cologne, I'm planning to call Grandpa and tell him. Tomorrow I'm going to see Mr. Bartenstein. Of course, I'm going to tell Mr. Schwarzfeld too, and Math-Gerhardt, and Miss

Venus, and my Conservatory professors that I'm now a physics student at the ETH in Zürich, where Einstein earned his Ph.D. and taught Physics.

Like all the universities in continental Europe, the ETH has no student dormitories. Everyone must find his or her place to live. Zürich is well known to be over-crowded. Last summer Saya contacted her best friend, Lisgi Brunner-Gyr, who in turn contacted Ruth Akert, a friend of hers, whose children had all grown up and left home.

When we arrive at the Akert house on the Zürichberg, it turns out that Mrs. Akert has been an avid reader of Saya's books. I get a cozy room under the roof with my own bath.

Saya wants to see the ETH. She eagerly goes from lecture hall to lecture hall, as if she herself would start studying math tomorrow. She likes the cafeteria. The food offered looks exceptionally good and the middle-aged cafeteria ladies remind her of the Penn-State cafeteria ladies, friendly and well nourished.

After Saya and Papa have left for Cologne, it sinks into me that from now on, I must plan everything by myself. Nobody will come to my room to wake me up in the morning. Nobody will wait for me with lunch or dinner.

As soon as the semester starts, I realize that lectures, seminars, lab works, and reading assignments leave me no time for shopping and cooking. I eat three meals a day at the cafeteria, which offers different menus, well-seasoned and tasty. As Saya pointed out, the chef is surely Italian or French.

Once a week Mrs. Akert invites me for dinner. This gives me a chance to meet Professor Konrad Akert, a prominent brain scientist and the founder of the Neurology Department in the School of Medicine at the University of Zürich. He likes to talk about brain waves, neurons, axons, cerebrospinal fluid, and encephalons—sometimes very explicitly, as if we were in an Anatomy Dissection lab with a corpse lying on the table.

Mrs. Akert sets the dinner table and shakes her head in disapproval. "Oh, Koni, please!" She begs him to change the subject, but he never does. With a twinkle in his eye he continues during dinner his graphic descriptions of brain anatomy. After a few times, I get accustomed to his ghastly stories and begin to inquire myself about the functions of the various parts of the brain.

Once I ask him if he could tell me where the soul lives.

"There is no place in the body for the soul to live, but maybe in the brain..."

"...but when death comes," I ask, "where does the soul go?"

He looks at me with his cynical, yet charming smile: "Mino, when you are dead, you are dead...that's it. As far as I'm concerned, just dead."

I think of Grandpa and his words that the soul lives on in the memory of the living.

Two years pass swiftly. I've never learned so much in such a short time, mostly math and its endless possibilities to help solving problems in science. Sometimes I feel that the beauty of math is like the beauty of Johann Sebastian Bach's music.

I joined the Alpine Club. All members happened to be men. Several of the Old Boys are now successful managers in Swiss banks. Some of them still practice climbing and lead junior teams. They have been to the top of the Matterhorn and Mont Blanc in different seasons, sometimes facing precarious situations. The best part of the Alpine Club is our monthly dinner in the Zunfthaus. Full-course, gorgeous, and, as a student, I don't have to pay.

Saya flies in from Cologne, because she has no patience

for sitting in the train for eight hours straight. When she comes, I dress up in my dark blue suit, which I have for special occasions like concerts or the opera. I wait for her at the airport, standing at the huge glass window, where I can see her coming from afar, beaming with joy and waving her hand as soon as she discovers me behind the glass. She stays for a week in a simple hotel near the Akerts' house, meeting with her literary agent, giving a lecture in a museum nearby, and reading from her books in a bookstore next to the Zunfthaus.

When I know Saya is coming here, I plan for a full evening program, a concert that has been recommended in the Neue Zürcher Zeitung, or a movie. Funny things happen with the movies. For some reason, the theaters where they play seem to change all the time. When we arrive at the place where a certain movie is supposed to be shown, we find out that it has been moved to a theater on the other side of the bridge. We run and run, sometimes for more than fifteen minutes, laughing and joking, until we get to the right place.

After the movies, we eat at the Mövenpick, talking about the plot, the characters, the scenes. Occasionally we are both so deeply moved by the film that we prefer to be silent. We walk through the old town to Saya's hotel on Zürichberg, but when we arrive there, she insists that she must take me to the Akert house, several blocks away. As soon as we arrive at the Akert house, I want to take her back to her hotel. So we go back and forth, until we both get tired enough to stop this pleasant ritual.

One nice thing about Zürich is the many concerts offered throughout the year, especially in the fall. Whenever I risk burning out from the intense work to succeed in the classes at ETH, I look for concerts. Lighthearted, Arthur Rubinstein playing Chopin, as he passed over the little mistakes he made, like letting the wind carry the music forth. After the performance I talked with him backstage, and he said I should come visit him in his home in Geneva. Vladimir Horowitz and Rudolph Serkin also gave piano concerts in Zürich. Both had a wonderful sense of humor. I spent a whole afternoon with Rudolph Serkin on the day after his concert and still remember how he put his hand on my shoulder when we departed and said, "I wish you fulfillment in all that you do in your life."

During one spring break I flew with Saya to New York where Joan Davis, her American literary agent, had arranged press events for her. We visited the Guggenheim. I could also have spent days in the Metropolitan Museum. We went to see Bob Fosse's Broadway show *Dancin'*, a fast, non-stop show that took my breath away.

Papa had to attend a conference in Richmond, Virginia, and he made his first stop in New York. We three meet with Joan Davis and her husband, Joe Kirchberger, in their condo on the top floor of 345 East 52nd Street. Both were originally from Berlin and had fled Germany during the Nazi time. Joan's father didn't make it and died in Auschwitz. Joe's parents perished in Dachau. His father was a mathematician with a Ph.D. from the University of Göttingen under the famous David Hilpert, whom many call the "Einstein of Mathematics." When I look at Joe from the side, his beautiful profile reminds me of a Greek sculpture. Joan could be Eleanor Roosevelt's sister. She has an infectious laugh.

Meeting Joan and Joe reminds me of my essay at the Kreuzgasse on anti-Semitism and its historical roots. How different it is to write about anti-Semitism in a high school class from sitting together with two wonderful people whose parents have been murdered as a result of this insane race science and the Aryan Paragraph. And Mr. Theil, my history teacher, didn't want to admit that anti-Semitism is deeply rooted in Christianity.

In the evening, we have dinner together at a cozy nearby Japanese restaurant. Joan and Saya start talking about some mischief they did together last fall in Frankfurt during the International Book Fair. Coming back late from two or three parties sponsored by some publishing houses, after more than a few sips of champagne, Saya had a sudden inspiration while tiptoeing through the long, silent hotel corridor. To the left and to the right, on the doorknobs, hotel guests had hung their room service breakfast orders—coffee or tea, orange juice, toast or rolls, boiled or scrambled eggs or sunny-side-up, sausage or ham. She started changing the numbers on the orders on one side of the corridor, room after room—five coffee, five tea, ten orange juice, seven toasts, three boiled and five scrambled eggs, seven sausages and twenty slices of ham. Hesitant at first, Joan started to scribble the breakfast orders on her side of the hallway, trying to suppress her giggle.

The following evening, again past midnight, the hotel corridor looked quiet as before, but a security guard was patrolling.

————

At the end of my first two years at the ETH, I pass the most demanding examination dreaded by all students. Everybody knows that after the first two years, the ETH mercilessly weeds out students who don't reach a preset standard. I am among the lucky ones who make it. Now I have earned my Bachelor's degree in Physics and Mathematics. One big step ahead.

Early September 1983 Saya flies back to Kyoto. Grandpa has asked her to come. Later that month I'm leaving with one of my friends from the Alpine Club to the Bernese Alps. Our goal is to climb the Schneehorn, one of the lesser peaks of the more than 4,100-meter-high Jungfrau massive. For the preparation of the trip we stay in a cottage that belongs to our Alpine Club.

We have to wait a few days for better weather. I spot through our cabin window a brand-new Mercedes-Benz, bright red, driving hesitantly along the narrow country road. The driver seems not to know where to go.

I look closer. It's Papa. I run out into the road.

"Did you buy a new car?" that's all I could ask.

"No, it's Mrs. Kaufmann's car. She let me have it so that I could drive fast."

"Why, why did you come?"

Papa takes me in his arms. "Grandpa passed away last night."

A SPECK FLOATING IN THE UNIVERSE

Saya arrives. I pick her up at the Zürich Airport and bring her to the hotel on the Zürichberg, where she always stays. She sleeps all day and all night. I'm concerned. I look into her room twice, but she is still fast asleep. She has travelled from Tokyo through Singapore and Bombay to Zürich, but this was not the first time she had come to Zürich via the long south route. Never before did she sleep non-stop for twenty-four hours. I call Papa in Cologne, but he says I should not wake her up.

Finally, late in the afternoon next day, I nonetheless decide to go again to her room. There she is, sitting on the side of her bed with a hollow look. "Minochan," she says. She glances at me and starts to sob. I have never seen her crying, so helpless, both arms hanging down, tears streaming over her face. I'm afraid of what might happen to her.

I pull tissue from her nightstand, wipe her tears and put my arms around her. "Saya-chan, Saya-chan, *nakanaide*!" I realize I'm doing exactly how she had comforted me whenever I was crying as a child.

"Minochan, *gomen ne*." She blows her nose again and again and shows a faint smile. She looks pale, no sparkle in her eyes. This is not the Saya I know. I'm frightened.

"Let's go have dinner," I suggest, trying to sound light-hearted, "I know a place." I take her to the Zunfthaus, where the Alpine Club members regularly dine. We get a window table. I order for both of us, since Saya is sitting there, disoriented, with no idea where she is.

Our dishes come, well prepared and tasty. I clean my plate in no time, because I'm very hungry. I have been working all day in my room in the Akert house, packing everything. I have to leave because the Akerts are going to move into a new smaller place in town, and I am scheduled to fly with Papa and Saya to Arizona. Papa has a research project at Arizona State University. I'll spend one year as a Research Assistant in the Department of Chemistry. We three would explore the Southwest of the U.S. on weekends.

Saya is eating very slowly, pausing after every bite. "I didn't realize," she says, "that Grandpa was going away. For the whole month, I was chatting with him, laughing and telling him what we all plan to do in the future. He was listening, quietly, saying few words. How could I know that he was going away? But he knew it...he knew it...he knew it."

I don't want her to start crying again. So I quickly say, "By being with Grandpa you made him happy. You would have made him sad, if you had started crying and wailing by his side."

"That's true, very true. After Grandpa died I discovered a poem, brushed on a *shikishi*, a farewell to all of us, written five days before his passing."

Saya can't finish her dinner, though she says several times that it was good and she liked it very much. I eat the rest including the two big chocolate mousses at the end. Since I've earned some extra money working in Professor Akert's neurology lab, I'm able to pay for this Zunfthaus dinner in full.

Walking back to her hotel, she is like floating with the image of her father in her mind. I only hope she doesn't start crying again. That's not the Saya I've known all my life.

———

Three of us start our autumn days in a two-bedroom condo in Tempe, Arizona. What weather! Flowers I've never seen before are blooming all over the gardens and at the campus corners. The air is crisp and cool at night, but during the day the sun is strong and bright. Papa has set up his experiments in the Physics Department and I'm doing Faraday and EPR measurements on insulating oxides in the Chemistry Department.

Every morning I leave home with Papa for work and Saya sits at her desk writing a new book, this time in Japanese. She tells me one copy of her latest German book, just published on time for the Frankfurter International Book Fair, reached her in Kyoto when she was with Grandpa. Included was a book review, three full pages long by Gerd Buzerius, who founded and owns *Die Zeit*. Saya translated the synopsis of the book and the review for Grandpa. He took the book in both hands, closed his eyes and told her she should write a book with the same content, but in Japanese, comparing Europe and Japan for the same historical period. That's what Saya has started to do now in Arizona.

But, typically for her, Saya would never be content with a rehash of her German book. She launches herself into a new research project, taking advantage of the treasures she has discovered in the university library. She spends days among the rows of bookshelves. She never expected to find here, in the middle of Arizona, thousands of Japanese books of academic quality and many Japanese lexica. I feel relieved that she no longer bursts into tears during dinner. She is coming out of her uncontrollable sadness. She is regaining her former self.

Once in a dream just before waking up in the morning, I saw Grandpa. He looked at me with his piercing but gentle eyes. Oh yes, I thought, he has not left us after all. I kept my dream to myself. I wasn't sure whether or not Saya would start crying again if I told her.

Christmas came, then the New Year, spring, and summer, when Saya finished her book, hundreds of pages, all written longhand. As she was sorting the pages, I told her about my dream. She smiled. "Of course, Minochan, Grandpa is with us and will always be."

As my employment in the Chemistry Department is half-time, I can spend the other half at the Department of Fine Arts, where I have been admitted as a graduate student. I learn how to use different stylographic pens to scratch images into wax-covered copper plates, to vary the width of the trace, and to learn the art of etching and of printing. I begin to appreciate how gifted the old masters were, in particular Albrecht Dürer. It would take a long climb to reach his level.

At the corner of the big studio in the Art Department I see a middle-aged man, white, with a beard. He always keeps to himself, never mixes with others or joins them in laughter. He goes by Michael, they tell me. He is a Vietnam veteran. He works on a huge canvas, drawing circles, endless circles in all colors and sizes. He reputedly has been coming for years, almost every day, producing circles, as if nothing else mattered in the world. During the Vietnam war, he lost his voice and became deaf. I look at his canvas at close range and plunge into a sudden spinning sensation. What is boiling in his mind? I want to talk with him, but he does not respond. He silently looks back at me with tender eyes. I think of Mr. Bartenstein.

I got my driver's license. It was easier than I had expected. For years, whenever I sat in Papa's car, on the passenger seat or in the back, I watched him driving. Sometimes I commented, "Not so fast," "Don't change lanes so often," "Keep more distance," or "Don't pass." Saya has no interest in cars or driving. She just looks at the passing scenery, pointing at a restaurant with a funny name, at small birds chasing a hawk, or at the picturesque evening sky. But she is extremely sensitive to bad driving, such as sudden acceleration or deceleration, jerking, or wavy driving. She quickly gets motion sickness and has to get out of the car.

According to Saya, Papa is an excellent driver. He has never had an accident, and he drives smoothly, no matter where or when. I was determined to drive like him and even better. With this background, I went to the DMV in Phoenix, passed both the paper test and the behind-the-wheel tests on the first try.

On the following weekend, I take Saya in Papa's car to Oak Creek Canyon, south of Flagstaff. Papa can't join us, as he has to look after his experiment. Saya and I enjoy the red rock cliffs, pine forests, and a small stream with opalescent water, the clean air and stillness of the place. Though we saw many parked cars and campers by the hundreds, the canyon swallows all human noise. As we drive back from our trip to the Oak Creek Canyon, Saya falls asleep in the passenger seat. This shows me how much she trusts my driving.

I take a one-week vacation to drive to the Grand Canyon. All alone. I duly register and then descend to the bottom of the canyon. I stay there for five days and five nights. The enormity of the place, the ever changing sunbeams falling on red, yellowish, and gray rocks, and the stillness and solitude have a calming effect on me, such as I have never before experienced. At night, the band of sky visible from the bottom of the Grand Canyon is filled with millions of stars. They seem to pull me up. Yes, I'm a speck floating in the universe.

GOLDEN HANDS

Checking out several rooms advertised in the local newspaper I've found a new place in Zürich. The room is clean and large, with windows facing east. I can use the family bathroom and kitchen. The room is furnished, and I can move in at once. It is very quiet, and I'll surely sleep deeply. It's on the other side of the lake, at the foot of the mountain, far from the inner city and far from the ETH, but all the necessary stores are in walking distance, and I can take the tram to the ETH. Mr. and Mrs. Arnold, the owners, seem to be kind and accommodating. All their children are out of the house and they are looking for a renter.

Mrs. Arnold seems to love polishing everything she can polish—door handles, kitchen sink, oven, refrigerator, kitchen table, floor, bathtub, bathroom floor, toilet, and the front porch. Everything is spotless and well maintained. I'm very cautious of not soiling anything. I think she also likes to do laundry, because she takes every piece away from me, washes it, hangs all my laundry in the basement, irons, folds it like new merchandise and places it back on my bed. She changes the linens once a week, making my bed like in a hotel.

One evening, when I come back from the ETH, I find a brand new upright piano in my room. Sometime ago I told Mrs. Arnold I used to play piano at home, but now I would concentrate on my work. She must have thought I would feel more at home in her house if a piano were standing in my room. I thank her and spend the next few hours on the piano. Since then she has started coming into my room to chat with me. After a few times, I'm running out of topics and I already know all about her children and grandchildren, and all her relatives and friends, and I know all of their names. I've explained to her that I need a few hours to finish my work before going to bed. I politely ask her to leave my room.

Mrs. Arnold must have been offended. On the next day,

Saya tried to reach me at the Arnold house, but I was still at the library. Mrs. Arnold answered phone and told Saya that I was a rotten scoundrel. Shocked and aghast, Saya asked her what I had done. Mrs. Arnold replied that my behavior was beyond description, unspeakable and atrocious. "He is out of anyone's control."

Saya tried to reach me in the Physics Department but didn't get through. She took the next flight to Zürich. When she arrived at the Arnold house I was just sitting in the kitchen and eating late breakfast. Both Arnolds were out, attending mass as they do three times a week. Saya is relieved that nothing has actually happened. Mrs. Arnold has had bouts of depression lately. It's all in her head.

Later in the day, when the Arnolds are back from church, Saya gives Mrs. Arnold a present, thanks her for taking care of my laundry, and for the beautiful piano. She praises her hand-made artworks hanging on every wall. All the time Mrs. Arnold is silent. We all end up sitting around the kitchen table, chatting about Switzerland, how it was when Mr. Arnold was a boy and how it is now. He vividly remembers his childhood in the 1910s, when only a few families in his village could afford shoes for their children. He tells us how tough the soles of his feet had become and how uncomfortable it felt for him wearing his first pair of shoes at the age of twelve. He was sent to town to work as a butcher's apprentice, passed the state examination, and became a butcher himself. Yet he had to change his job because he hurt his shoulder carrying half a cow on his back. He became a tram-driver, drove for many decades all over Zürich and retired some years ago.

Saya likes Mr. Arnold and the little twinkle in his eyes. He has a fine sense of humor. "He is one of very few people I've ever met," she tells me afterward, "without any inferiority or superiority complex. He is at peace with himself and with the world. I want to listen to him more next time; the way he talks is captivating." She also likes Mrs. Arnold for her goodness of heart.

After Saya's visit Mrs. Arnold stops coming to my room to talk, though Saya, during her visit, never brought up the incident of Mrs. Arnold's strange telephone call.

As graduate student, I have to start looking for a professor with whom I would like to do my thesis. At the same time, I have to take a number of advanced physics classes, which I find interesting and easy—easy because I've built up a strong background in math. Nothing scares me.

I'm taking courses in nuclear physics and quantum mechanics. Next semester it will be solid-state physics and electrodynamics. I marvel at the beauty of the Maxwell equations. I'm toying with the idea of going into theoretical physics, but Professor Rice advises me to focus on experimental physics. "You have golden hands, Mino," he says with an unmistakably Irish accent. "I watched you in the Graduate Lab. You built things beautifuly, and, importantly, they always worked."

Rice is professor for theoretical physics. Coming from the Bell Labs, he is interested in materials with charge and spin density waves and on metal-insulator transitions, something Papa has also mentioned.

"Am I not good enough in theoretical physics?" I ask him straight out.

"You are good in theory," he answers with a broad grin. "You may be as good as theorists get, but good experimentalists—they are scarce. That's what we need."

Now I understand. Professor Rise might have had a hand in the decision to assign me to the Graduate Physics Lab. I'm now in charge of it. Every Wednesday afternoon. It's a big thing, twenty students, many older than me. I have to set up experiments for them, make sure that all parts are running, and oversee the whole group.

The department has asked me whether we need to replace some of the electronic equipment. Things might have become outdated. I suggest that they buy some of the latest Hewlett-Packard multimeters and the fastest multi-channel Keithley oscilloscopes. I can use them to update some of the old experiments. The lab is great fun and I often stay late, after all students have left, sometimes to midnight and into the early morning hours.

It's 1985 and Papa has started at the NASA Ames Research Center as a Senior National Research Council Fellow. I spend the summer months in California, staying with Saya and Papa, working as an International Fellow at the Stanford Research Institute, in their Molecular Physics Lab in Menlo Park. Saya often uses the library at the Hoover Institution, Stanford University, for her next project. We live in a rented condo in Palo Alto. My work on surface physics and surface analysis is quite enthralling.

On the weekend, I drive with Saya to Half Moon Bay, run around on the beach, watch hundreds of birds flying in constantly changing formations. At sundown, we watch the orange disk of the sun dipping into the ocean. During the week, we often go to University Avenue to eat sushi, prepared by Toshi, a sushi master trained in Japan. We spend hours in the Stanford bookstore and sometimes I go to NASA with Papa. Once I drive with Saya to the Aquarium in Monterey Bay. We see dolphins close to the coast, playing in the water.

Returning to Zürich for the fall semester, I attend an ETH seminar given by Dr. Heinrich Rohrer of IBM near Zürich. He talks about the Scanning Tunneling Microscope (STM) he has built. He shows images of individual atoms that he has obtained with his instrument. I'm blown away. I decide on the spot to build one STM by myself. I read Rohrer's publications and his technical reports with the circuit. Pretty intricate. I start putting together the components. The evenings are quiet in the lab without interruption.

In April 1986, Papa comes to visit me from California. I take him to the Graduate Physics Lab, where I've stored in a locked cabinet my toy, my own Scanning Tunneling Microscope with its delicate mechanical parts that must be stable enough to image individual atoms at a resolution of better than an angstrom, a millionth of the width of a hair, and the electronics that rely on complex feedback loops to translate tiny electric currents into images.

I ask Papa to sit down while I zero in the scanning tip onto my test surface, a graphite single crystal. The computer screen flickers, but slowly, after a few trials, the hexagons of the graphite structure become visible. The six carbon atoms that make up the graphite ring appear with impressive clarity. Papa is speechless. When he comes to his senses, he takes me in his arms, muttering, "Mino, Mino, marvelous, great, splendid!"

In the fall of the same year, 1986, Heinrich Rohrer and his coworker Gerd Binning receive the Nobel Prize in physics for their Scanning Tunneling Microscope. I later hear from Heini Rohrer that the STM that I built and donated to the ETH Graduate Physics Lab was the first STM ever built outside IBM.

HMM...UMM

Jørg L. Olsen is from Denmark. He grew up in Copenhagen, was educated in Oxford and has been a tenured professor of physics at ETH for decades. He specializes in Low Temperature Physics. Quite a large number of his Ph.D. students have become professors all over Europe, and several of them won Nobel Prizes, including Heinrich Rohrer of IBM, the inventor of the STM.

I like the way Professor Olsen teaches physics. Actually,

he doesn't teach physics systematically, but often starts class with a question that he throws at the students. He tells anecdotes, seemingly unrelated, makes comments in passing and takes the students to an *aha!* moment that casts light on the original question. His teaching style is rather tricky. It's easy to be deceived by his almost conversational tone, until he inserts, in passing, some really high-level mathematics. I realize I have to be very attentive all the time so as not to miss things he throws in casually. I have to equip myself with all the knowledge of the books and published articles he assigns.

One late afternoon in the Graduate Lab, while I was transferring liquid helium to push the temperature of my sample down into the milliKelvin range, Professor Olsen came in and sat at the corner of the work bench. I was too busy with my experiment to talk to him. He obviously noticed it. "Take your time, Mino."

After I was done with the liquid helium transfer and had adjusted the control instrument, Professor Olsen started to talk in his hesitant Oxford style with lots of interspersed *hmm umm*s. "Do you want to stay in low temperature physics?" he asked.

I was surprised. "Yes."

"*Hmm umm* why?"

"Superconductivity is dazzling. It changes so many laws of physics and there is still much to be discovered."

I was wondering whether this would be a good moment to ask about a Ph.D. thesis. I was still in the master's program and had not yet taken the Ph.D. qualification exams.

Professor Olsen must have read my mind: "*Hmm, umm,* well, in that case you could do your Ph.D. with me."

I felt like jumping up and hugging him, but of course I didn't. I only said, "Thank you, Professor Olsen."

He stood up and came over to my side of the bench. "All right then, Mino." He put his hand on my shoulder.

Papa has been shuttling between NASA Ames in California and University of Cologne in Germany every month, and Saya is flying frequently between California, Japan, and Cologne. One of them always comes to visit me for a day or two or for a week at most. With Papa, I spend the time mostly in the lab, talking physics.

With Saya, I go to the movies or to a concert. We eat dinner, talking and laughing so much that it makes us hungry again. Saya meets Professor Olsen and it turns out that he has been an avid reader of her books. He talks about East Asia, where he spent a few years of his childhood in Kobe, Shanghai, Hong Kong, and India as the son of a Danish engineer on diplomatic assignments. One of Saya's novels takes place in Japan before WWII. Several scenes remind him, he says, of the time when he was a young boy roaming the streets of Kobe or Shanghai.

MOON RISING OVER MOUNT HIEI

In the summer of 1987, the ETH sends me to Kyoto to participate in the International Superconductivity Conference. Though I'm only a Ph.D. candidate in Low Temperature Physics, I'm one of the official delegates. Professor Olsen must have arranged it long ago with the ETH administration. I'm representing him, so to speak. The happiest news is that Saya joins me in Kyoto. She is in Hawaii as Writer-in-Residence at the East-West Center, and she decided to be with me in Kyoto. We stay in the Kyoto Hotel, where most international conference participants are housed.

Saya stays in the hotel finishing her next book in Japanese. It's almost done, she says, but then unforeseen changes and a number of last-minute ideas must be incorporated.

Every morning a luxurious bus with air conditioning in full blast comes to pick up the participants at the front of the hotel. It takes us to the International Conference Hall in the northeastern part of the Kyoto valley, where the Eastern mountain range begins. The Conference Hall, surrounded by a large park, has everything necessary for smooth proceedings. The only drawback is the air conditioning in full blast, which brings the temperature down drastically. Many participants from Europe are not accustomed to such an icebox. I'm one of them. We are freezing.

Dr. Alex Müller and Dr. Georg Bednorz, both from IBM-Zürich, are the center of attention. Everybody is talking about their discovery of high-temperature superconductivity in ceramic materials. Both are surrounded everywhere by journalists, photographers, and TV cameras.

On the last evening, all Conference participants get together for the farewell dinner. Japanese, Chinese, and French delicacies abound. I've never seen such varieties of food at one place. It is, as if a Japanese hot springs inn, the Zunfthaus in Zürich, and the Chinese Palace in San Francisco had all come to Kyoto with their food. All Conference participants are busy gobbling down what they can. Some are clearly overdoing the drinking.

Saya asks me what Müller and Bednorz have discovered, since they are constantly surrounded by cameramen and can barely eat or drink. I tell her they have shown that certain copper oxides become superconducting at a temperature much higher than anybody thought would be possible. Saya nods but I can see in her eyes that her thoughts have already wandered away.

Quite remarkable is what happened prior to the Müller and Bednorz publication. They had submitted their paper for peer review to a series of internationally renowned journals, but all turned them down, saying that their "discovery" was without substance or merit. Neither scientist gave up. Finally, Müller and Bednorz succeeded in publishing their paper in the venerable *Zeitschrift der Physik,* which *The New York*

Times called an "obscure journal," though it was where Einstein published his benchmark papers in his miracle year of 1905. Within weeks of the publication of the Müller-Bednorz paper, it became a world-wide sensation, after several physicists had quickly verified and confirmed its findings.

The reception hall is echoing with voices, glasses, knives, and forks. Saya can't make much sense of what she hears about the discovery. She nods with admiration in her eyes for my knowledge. I tell her I know both, especially Dr. Müller. She says, "I'd like to meet him."

I take her to Dr. Müller. After a few introductory words, Dr. Müller mentions that before this trip to Japan, he had read a remarkable book, in fact a bestseller, which *Die Zeit* had reviewed using three full pages. "It provided me," he said, "with some deep insights into European mentality in relation to Japan...a very different way of looking at history. ...I don't remember the author's name...a Japanese name."

I turn to Dr. Müller. "The author stands in front of you...," I say. "My mother," with emphasis on *"my"*.

Startled, Dr. Müller shakes Saya's hands once more. "What a surprise."

A day after the conference, I come down with a fever, a headache and a sore throat. I feel miserable. Saya arranges with the hotel manager to extend our stay for five more days. She thinks it's all due to the relentless onslaught of chill and draft. The hotel manager mentions that five more guests, all Europeans, also got sick. According to Saya, Japanese are overly proud of their air conditioning, so they over-use it everywhere, just as Americans used to make people freeze from spring to fall in air-conditioned office buildings, restaurants, trains, busses, and streetcars.

Saya thinks I should take a hot bath and then sleep as long as I can. She keeps our room air-condition free, though it's the end of August and hot outside. I sleep all night through. Next morning, I wake up very hungry. Saya orders a big breakfast to the room. My fever is slightly down, but the headache and sore throat linger on. Saya starts to get exactly the same symptoms: fever, sore throat, and a general run-down feeling. We sleep most of the time and order our meals brought to the room. The hotel room service crew kindly adjusts to our rhythm. Finally, after a few days, we recover.

The following evening, we go to the Daitokuji Zen Temple, where I had made drawings years ago. We walk in the large precinct under pine trees and go through the gate deep into the temple garden to Grandpa's grave. The cicadas are singing, and the moon is rising over Mount Hiei. We see the fresh tombstone on white gravel.

Facing the headstone with Grandpa's name carved in granite, I think of the time when he told me in the East Room of the Kenkun Shrine about his father's sudden passing and about his lifelong quest for the hereafter. He studied Buddhist sutras, visited Buddhist temples, Christian churches, read the Old and New Testaments, meditated in the forests, found some important pieces of truth in the I-Ching. He became a Shinto priest because this gave him the freedom to study further and further and further. After years of search, he became proficient in the multi-layered combinatorics that is the essence of the I-Ching. An endless stream of people came to see him, as if on a pilgrimage, to benefit from his wisdom.

Now, standing in front of his tombstone, I ask myself where Grandpa has gone. Saya carries a bucketful of water from a nearby well, scoops it up with a bamboo dipper and pours it gently over the stone. She squats down in front of it and starts talking to Grandpa. She tells him all about what we have done in Kyoto, as if he were listening to her.

Then she takes my hand and starts to sing, *"Sho sho sho-joji shojoji no niwa wa...."* Grandpa used to sing the song for Saya when she was a little girl and she always sang for me when I was small. Now we sing together for him while the moon is rising over Mount Hiei.

After I return to the ETH to continue my Ph.D. work, we hear that Alex Müller and Georg Bednorz have been awarded the Nobel Prize in Physics for their discovery of the high Tc superconducting copper oxide.

DAMN PAPER

Professor Olsen has asked me to write a paper based on my work so far. I write a draft in one sitting, eight pages long, and show it to him. He reads in silence, shakes his head and returns the pages to me without comment. I go back to work on the text for the rest of the afternoon and bring it back to him.

He goes through the pages very attentively, shakes his head on page three and on page five. He hands the whole stack back to me without comment.

I write and rewrite the whole thing all evening. Hearing his door closing, I run and catch him before he leaves. I hand him my newly revised version.

Professor Olsen goes back into his office, sits down at his desk, and starts reading again, muttering "hmm, *hmm... umm, umm.*" He stops on page six and gives me back all pages without any further word. He leaves for home.

I'm infuriated with Professor Olsen. For the first time. Why doesn't he point out to me what's wrong with my paper? He could have done it already with the first version. Why can't he tell me which lines, which words, or which equations in my paper are bothering him? Coming back to my room, I read and reread the damn paper. Everything looks fine to me; fabulous, great, even perfect. I feel hungry and realize that I skipped lunch and supper.

While eating at home slices of bread with smoked salmon and salad, I'm thinking of my paper. By now I know it by heart. I have become convinced that there is something insane about Professor Olsen. Definitely. My paper presents all that needs to be described and does so in a concise, proper, persuasive language.

Exhausted I fall asleep.

Waking up in the middle of the night, an entirely different approach to my paper comes to my mind, based on the same experiment and the same content. I jump out of bed, rework the text, reduce the number of pages from eight to six, eliminate superfluous words and long winding arguments.

In the morning, I hand my new version to Professor Olsen, who reads it with visible concentration. He breathes deeply at the end and put his hand on my shoulder: "Mino, you've got it."

"Thank you." I try not to put my arms around him, as I would have done with Papa or Saya.

Now I appreciate the way Professor Olsen is dealing with me. It was a painstaking procedure for him and for me. At the same time, it was the best education, and probably the only way to let me find out by myself how to write a good scientific paper. If Professor Olsen had pointed out right from the beginning what was wrong and what should be done differently, I would have accepted his judgment and changed my paper accordingly. After all, he is an experienced, Oxford-trained physicist, a professor, a writer and educator who has mentored several Nobel Prize winners among his Ph.D. students.

But, so what? I would never have discovered by myself what was lacking in me and in my text. I would never have developed the skill and intuition to write a good scientific paper.

Over the next three years, while I'm doing my experiments and as I begin writing my Ph.D. thesis, the same process happens occasionally. Professor Olsen never says what he doesn't like with a chapter I'm showing him. He shakes his head and gives it back to me. After contemplating for hours or days, sometimes for a week, I figure it out. Being a perfectionist, Professor Olsen never compromises halfway. I learn to watch his posture while he's reading, to note subtle expressions in his eyes and to catch where he utters *"hmm, hmm,"*

Gradually I learn to write a chapter or a new paper that he accepts in the first round. It is an invaluable lesson.

From time to time I take the train to Basel, only one hour away, where Dr. Hilti of Ciba-Geigy is preparing in his research lab the crystals, delicate gold-organic compounds, which become superconducting when cooled to liquid helium temperatures. These single crystals are difficult to grow, but I need them, if possible in millimeter size, to measure their electric resistance, which vanishes completely when I cool them down.

One day at lunch, Professor Olsen says to me casually, "Mino, why don't you call me just Jørg? That saves a lot of time."

"All right, Jørg," I reply, feeling honored and happy.

Besides my project for Ph.D., I'm taking classes in computer science and listening to lectures on cosmology. I learn about stars and galaxies, about black holes and the Big Bang. The Big Bang theory is controversial, still not quite settled. Distinguished speakers from around the world give seminars at the ETH, in addition to the four physicists from the IBM lab near Zürich, who won Nobel prizes, all closely associated with our ETH Physics department.

I continue helping in the Graduate Lab for the younger physics students and help them use properly the Scanning Tunneling Microscope, which I had built before Heini Rohrer and Gerd Binning received their Nobel Prize.

Whenever I see my instrument in operation and see the shapes of individual atoms appearing on the computer screen,

I am humbled by the thought that my universe spans the whole gamut of dimensions from the smallest of small individual atoms, to the thousands and millions of light years that separate our Milky Way galaxy from other galaxies, and the billions of light years that separate us from the Big Bang.

The copper oxide materials that Alex Müller and Georg Bednorz have studied in such great detail become superconductive at a much higher temperature than my delicate gold-organic crystals. Their discovery was sensational and certainly deserved a Nobel Prize. However, even though hundreds of the best physicists around the world immediately began working on this new class of materials, nobody seems to have an explanation for why they become superconductive.

When Alex Müller spoke at a workshop in Interlaken, we students surrounded him afterwards, asking question after question for half the night: "Is it possible that the electrons in the copper oxide materials remain in the superconductive state?" "Spin-coupled in pairs?" "Why does spin coupling allow electric current to flow without any resistance?"

I think about the experiments that Papa had conducted in his lab at the University of Cologne before going to the USA. Studying the simplest of all oxide materials, magnesium oxide, he had discovered that they contain certain imperfections on the atomic scale, called peroxy defects. When Papa breaks these peroxy defects, they form electronic states that remain spin coupled at even higher temperatures, much higher than the superconductive copper oxide materials.

The main difference is that in Papa's case, these are not electrons, but sites in the atomic structure where an electron is missing, so-called defect electrons or holes. This creates a challenging situation that requires quantum mechanics. Before leaving for America, Papa asked me for help, and we spent many hours discussing the best way to address the unusual spin coupling that he had discovered. We started to write a paper, which we would send off for publication in the reputable *Journal of Solid State Physics*.

When I want to free myself from physics, I take a train to Zug, where Lisgi Brunner-Gyr, Saya's best friend in Switzerland, welcomes me in her elegant house overlooking the lake. She is a tall, humorous, elderly lady with five grown children who have left the house. Emma, her housekeeper, always cooks her special *eglifilet* and her gardener takes care of the fruits in her garden.

Once Lisgi takes me to her mother, who resides by herself in a very old house with precious oil paintings hanging on the walls. The mother is a tiny old lady with smiling eyes. One night, a few years back, burglars broke into her house. The old lady woke up and surprised them as they were busy taking paintings down from the walls. She offered them coffee and entertained them with tales of Indonesia, where she had spent her youth with her engineer father and painter mother. The tiny old lady enchanted the burglars so thoroughly that they put the paintings back on the walls. They thanked her for the coffee and her hospitality, then left the house, as if they had been her honorable guests.

WILD GENTIANS

My paper with Papa titled "Highly Mobile Oxygen Holes in Magnesium Oxide" was published in the prestigious journal *Physical Review Letters* in 1989. We are co-authors. But, to tell the truth, when we started, Papa had written the entire text. I was not happy with the way he presented it all. I felt like Jørg Olsen must have felt reading my first draft. I could have uttered *Hum hmm hmm*, shaken my head, and returned the pages to Papa without a word. Never would he have sensed what's wrong with the paper, let alone understood that I was not pleased with it. After all, he had already written many publications, and for him this was just another one of them.

But not for me. If I joined him as co-author, we must get it right.

I sat down, ripped Papa's text apart, discarded several sentences, changed some adjectives, wrote a few new paragraphs, and put them together in my way—the way I learned from Jørg.

Papa was stupefied. Speechless. After a while he said, "Well done, Mino!" And so we sent it out. This would be the beginning of a long list of joint publications.

I keep track of the new developments in the personal computer arena. There is the Atari 400/800 and the Commodore VIC, but the one I like best is the Apple. Last time I was in California, Papa asked me which home computer he should buy. We looked at different possibilities but decided on an Apple Macintosh. It comes with a keyboard and a mouse. It has a very intuitive operating system that even Papa will find not too difficult to use.

Back at the ETH, some of the Physics and Engineering students start buying Macintoshs, and the word spreads that I've become sort of an Apple expert. I know the Apple operating system in and out, and I know all the shortcuts you can take by typing simple commands on the keyboard. Almost every day at the ETH somebody comes and needs help. After a month, I have to set up office hours to manage the steady stream of Apple aficionados.

The word even spreads to the University of Zürich. One day a girl comes to see me, lugging her new Macintosh to my office on the third floor of the Physics Building. The screen on her computer is frozen. I'm able to help her on the spot.

Three days later she comes back with the same problem. After I show her how to unfreeze her screen, we chat for a while. She is Swiss, studying biology. Amelia is her name. She is just starting her master's research on gentians, the large, trumpet-shaped flowers that grow in the Alps, intensely blue, azure, like the sky in California. I have seen gentians during my trips with the friends from the Alpine Club when we scaled some moderately high peaks in the Berner Oberland.

Amelia tells me that gentians are becoming rare because so many people dig out the roots to brew bitter drinks. Indeed, since ancient times gentian roots have been used to prepare medicinal tinctures and to flavor apéritifs. Though it is forbidden by law, people keep digging them out. In some parts of Switzerland there are no gentians left in the wild. Amelia's master's thesis is about this problem. Her first task is to find out where in the Berner Oberland gentians used to be abundant and, second, what could be done to reintroduce them in the wild.

I like Amelia—her gentle way of talking, her melodious Swiss intonation, her affection for the gentians, and her way of telling me about her project. When she comes back for the third time, she no longer brings her Macintosh along. I tell her where I saw gentians during one of my outings in the Berner Oberland and suggest we should go climbing the coming Sunday, weather permitting. Amelia looks surprised and happy.

We meet at five in the morning at the Zürich train station. Amelia has put her chestnut brown hair up into a ponytail. She wears an olive-green anorak and mountaineering boots and carries a backpack. We take the express train to Bern, where we change to a local train and then to a bus. To reach the valley where I saw gentians in bloom last spring, we hike more than two hours up the mountain slope to the trail head. There we put on serious mountain gear.

I have carried everything we need in my backpack, including my well-used harness and a harness for her, borrowed from our Alpine Club. Also, twenty meters of rope and quickdraws to connect the rope to the bolt anchors, which I would hammer into cracks in the rock face. This will allow the rope to move smoothly through the anchors, so that I can secure Amelia while she traverses the steepest part of the ascent. Amelia has never worn a harness or used quick-draws with a rope. She is excited and faithfully follows my instructions.

All goes smoothly, and we reach the top of the wall around noon. Before us a valley with lush grass growing on the flat floor and gentle rocky slopes on both sides. It's a hanging valley, I tell Amelia, cut by a glacier during the Ice Age, when the entire Alps were covered with thick ice. This valley below was carved out by a large glacier, our side valley by a small glacier that had joined the large one. When the ice melted at the end of the Ice Age, this side valley became a hanging valley, separated from the big valley by the steep wall we had to climb through.

"That's why this valley is difficult to reach, and few people come up here. That's why there are still a lot of wild gentians here."

As we cross the valley floor, moving up the increasingly steep slope on the sunny side, we reach many patches of gentians, azure blue, growing between the rocks wherever there is a bit of soil to support them.

Amelia is getting more and more excited.

"Never seen so many at one place," she exclaims with joy. "What a beautiful place. Since it's a hanging valley, people can't reach it easily."

Amelia counts the number of flowers per square meter, sketches their locations in her notebook, takes photographs, and gathers a dozen or more soil samples to bring back to her lab.

Back in Zürich, she comes every week, bringing sand-wiches, orange juice, and black chocolate after she has found out that chocolate is my favorite dessert. She brings the first chapter of her thesis and reads it to me with her melodious Swiss intonation. She doesn't want to show me the pages, because, as she says, they are still muddled up.

She asks me why I'm cleaning up my room. I tell her that I'll be leaving after my Ph.D. defense in the fall.

"...So soon? Where are you going?"

"Don't know yet. Maybe England or Germany, maybe America." I tell her that Dr. Needham has recommended me for Cambridge and Professor Olsen wants me to go to Oxford. Also, one of the Max Planck Institutes and the University of München have offered me a post-doc position.

Amelia looks sad. "Why so soon? You should stay here, somewhere in Switzerland. Why America?"

"I have several offers. Actually, I'm American, though I've never lived there for any long stretches of time."

She nods. "But so far away. So soon...."

I tell her I would surely come back to Switzerland sometime, and she should come to visit me in America. "We'll be in touch."

"So far away. So soon..."

I suggest that we climb once more the same hanging valley in the Berner Oberland. "What about next Sunday? We might find gentians still in bloom."

Amelia smiles. A dimple darts on her left cheek.

It is a clear day in early fall. Like last time, we take the train and bus and hike for hours. I put on my harness and help Amelia into hers. We work with the rope through the steepest part. Suddenly surrounded again by so many gentians, azure blue. Enveloped in the breathtaking scenery, we just stand there holding hands. I take her into my arms and kiss her. She starts to cry.

"When you finish your MS degree, you should come to visit me," I say.

Amelia doesn't reply. Fighting back her tears, she only murmurs, "Please stay in Switzerland...don't go away."

LEAVING

The day of my oral defense is approaching. Saya and Papa have come from San Francisco. They buy a few bottles of French champagne and a dozen champagne glasses in anticipation of a celebration after the defense. Mrs. Arnold has ironed my white shirt with starch, so that I'd look official with my dark blue suit and a necktie that Saya has bought for the occasion.

The defense takes place in one of the smaller lecture halls of our Physics Department. Jørg Olsen and Alex Müller sit in the front row, together with Bruno Hilti from Ciba-Geigy and the Dean of the School of Science. Saya and Papa are in the last row on the right and a bunch of my friends including, Jan Korvink, are on the left. I feel very relaxed, though I have to pack my two hundred-page-long thesis into thirty minutes.

Saya is the only one in the room who nods and smiles. Whenever I look in her direction, she is beaming. I know that not a single word makes sense to her, but she is probably evaluating my presentation from the point of view of breathing, pronunciation, intonation, timing, eye contact, movement of my hands, and facial expression.

Papa looks stern, as usual, sitting tight and unyielding, as if he were the one being examined.

My four judges on the first row are intently listening, taking notes. The lecture hall is eerily quiet except for my voice echoing.

After my talk, questions come from Alex Müller and Bruno Hilti, tough questions. Professor Olsen is smiling, and the Dean is quiet.

Then everybody is asked to leave the room, to wait in
suspense while the four judges decide pass or fail. We go out
of the room and close the door.

Saya has barely time to say, "Great, Minochan!" before
the door opens again, and the Dean appears. "Congratula-
tions! Quite remarkable!" He shakes my hand.

We go back into the lecture room. Saya distributes glasses
to everybody; then Papa opens the champagne bottles and
fills every glass.

I go to Professor Olsen. "Thank you, Jørg." He gives me
a big hug. We both try to hide a few tears.

I have invited Amelia to join us for dinner at the Au Premier. Saya and Papa like her very much.

Next morning, we drive to Marburg, eight hours straight,
to see Hans and Käthe Hooss, who took care of Saya and me
during our first winter in Germany, because we had nowhere
else to live in Göttingen. I was a baby then. I saw the Hoosses
a few times while growing up, but it has been years since
my last visit. Hans and Käthe wait for us in their living
room, the table set with Käthe's home-baked pie and can-
dles. Upstairs Opa and Oma Hooss are no longer alive. We
talk about their golden wedding anniversary a quarter century ago, laugh and chat for a few hours, and then leave for
Cologne.

Staying in our house with our renters, a British couple, I
go to see Mr. Schwarzfeld across the street. He has become
frail and sickly. No strength left to paint. I don't know what
to say, because I never imagined him to be without passion
for painting and for life. A young man from the City Social
Services comes in and calls out loud. "Hey, old man, have
you peed in your pants again?"

Mr. Schwarzfeld stands in the corridor like a shadow.

"Listen!" I say to the young man, "that's not the way to
talk to a great painter and great teacher. You'd better watch

your tongue." The Social Services guy responds with a scornful grimace.

I take Mr. Schwarzfeld's hands. "You have done so much for me. I'll never forget, and I'll continue to draw." I was too sad to say goodbye, though I knew it was the last time I would see him.

Miss Venus comes all the way from Göttingen to see us in Cologne. We play together on our Bechstein piano, which is ridiculously out of tune. She is as cheerful a teacher as ever. We have a dinner together at a Chinese restaurant. We talk about my first concert at the age of six, when I forgot my music score and instead produced my own improvised melodies on the piano. Miss Venus remembers the incident and still laughs about it.

I visit Mr. Bartenstein. He has recovered and lives by himself in his old apartment. His room is full of bookshelves, on three sides, from the floor to the ceiling, and open books are scattered all around. He does not want to talk about Kreuzgasse and the time when he was our most inspiring teacher. A connoisseur of French literature, who has shown me what poetry is and can be, who has embraced me for having Japanese heritage. He is sitting among his many open books, a shadow of his former self. Surprisingly, however, he remembers that I came to the psychiatric ward every afternoon for three months, taking him out, if the weather was fine, on his wheelchair to the park. Bartenstein remembers details of those afternoons, even small details, where we went and which questions I asked. With a faint smile he sets out to answer my questions from almost ten years ago. It is as if the clock had stopped.

Back in Zürich, we sit with the Arnolds at their kitchen table, sipping champagne. Next day we invite Alex Müller and his wife for dinner in the Zunfthaus. Dr. and Mrs. Rice invite us for dinner at their home. We also invite the Akerts, with whom I stayed for the first two years of my ETH days. We invite Jørg and his wife, and we drive to Zug to see Lisgi. She invites us for dinner and serves, as usual, *eglifilet*.

"Mino, have you met any nice girls?" Lisgi asks. "Anyone special?"

"Yes, Amelia, a biology student. She is sincere and pretty, too. I like her very much."

"Where is she from?"

"From St. Gallen."

"Sooo, she's Swiss. That's a problem."

Lisgi tells me how very conservative the Swiss are, in general. No Swiss parents want their daughters to marry non-Swiss men. That's just not done. They even make a fuss if a daughter wants to marry a man from the next valley. If a Zürich man wants to marry a Bern girl, that's a problem. Sure, it's less of a problem now than fifty years ago, but it's still a problem. If a Swiss girl marries someone from outside the country, that man shouldn't expect her to follow him. He better comes to Switzerland or else....

I think of Jørg Olsen. The son of a Danish diplomat, grew up in Denmark, Japan, and China, and was educated at Oxford. When he married his Zürich wife, he moved to Zürich. Professor Rice from Ireland had to move to Zürich too, after he had met his wife in America. She is Swiss. The list goes on and on.

Now I understand why Amelia said, "Please stay in Switzerland."

Like most parents back in the 1960s and '70s, my mama and papa took lots of photographs, and since I was their only child, most of their pictures were of me. Here are a few of their favorites, pictures that also recall memorable moments in my younger years.

With my baby elephant in our tatami room in Göttingen
at the age of one year and two months.

Singing with Mama about racoons dancing in the
temple garden at night. Age three.

At the age of four, I was the youngest guest at the party in Marburg celebrating the golden wedding anniversary of Oma and Opa Hooss. I ate everything on my plate—and a second piece of cake.

After having been beaten up in Cologne by German kids
and called a "dirty Chink," I am happy to be at the
École Chaperon Rouge in Crans-sur-Sierre,
Switzerland, at age six.

Practicing at my Bechstein grand piano at home at age sixteen.

Freshly minted graduate with abitur and baccalaureate
in Cologne at age nineteen.

As I grew, I also developed a passion for pictures, but instead of pointing a camera, I found that I "saw" much better with the help of pens and pencils and other hand-held tools. The rest of this picture album is a sampling of sights I wanted never to forget.

A medieval tower in Cologne, which I drew during my first artist's trip with Mr. Schwarzfeld, starting to learn perspective.

Entrance to the Schwab Auditorium on the campus of
Penn State University, where Saya studied Theater Arts.

Palm trees waving in the wind under the bright Florida sun.

Within the drawing: "Notre-Dame de Paris"
2/4/80   Mino

Fascinated with lights and shadows in Notre Dame de Paris.

Sorrow emanating from two shadows in Notre Dame de Paris.

A narrow street in an old part of Paris on an early summer morning.

Familiar view of the Gross Münster in Zürich, my home town during university years,
seen across the Limmat just before it empties into the Lake of Zürich.

Grandpa's Shinto shrine, Kenkun Jinja, on the tree-covered Funaoka Hill,
which is surrounded on all sides by the city of Kyoto.

The entrance to an old residence in the imperial city of Kyoto caught
my attention because of the seemingly random, but in truth
intentional, asymmetry of the bridge design.

A narrow alley in the medieval part of Cologne,
rebuilt after the total destruction caused by WWII.

COSMOLOGY

FIRP

Since I was little, I have been fascinated by the night sky so full of stars. At the Chaperon Rouge in Switzerland, at the age of six, I often sneaked out of my room at night, went into the backyard, away from the building, and looked at the sky. Later in Cologne, I went with the Computer Club to the Effelsberg Radio Telescope. There I saw for the first time stars that are embedded in the gigantic dust clouds that fill the plane of our Milky Way galaxy. I saw the birth of stars inside dust clouds. Later, lying on my back at the bottom of the Grand Canyon, taking in the night sky, I saw millions of stars so close it seemed as if I could reach out to them and they would carry me through the universe.

At the ETH, while I was still working on my Ph.D., Jørg Olsen told me about a planned NASA-Japan satellite mission. It would have an instrument onboard, he said, designed to measure the extremely faint radiation emitted by the cold dust in the far-away interstellar and extragalactic space. He gave me some of the specifications, and we talked at length about how difficult it would be to build an instrument that would work in the zero-gravity environment of space, recording images at this faint, very long wavelength infrared radiation. "SubKelvin temperatures," Jørg said, with a scrutinizing glance at my face. "That's what will be needed. If you can pull it off, these would be images of something that has never been seen before." He seemed intent on nudging me towards getting interested in this technological challenge.

I tried to learn more about the IRTS mission, the Infrared Telescope in Space. The instrument Jørg had mentioned, was FIRP, the Far Infrared Photometer. According to the information I could gather, the design called for a parabolic mirror to be cooled to liquid helium temperature and a core, the bolometer array, to be cooled to milliKelvin temperatures, as close as possible to absolute zero. Something like this has never been done in space.

Now I've received the offer to join as a post-doc the Astrophysics group in the Physics and Astronomy Department of the University of California in Berkeley. The job description mentions the IRTS as a joint NASA-Japan satellite project, and specifically the FIRP. If I take the job, I would be responsible for the completion of the FIRP, its calibration and integration onto the satellite platform.

I fly to San Francisco to meet the leader of the U.S. team for the IRTS, Dr. Andrew Lange, a young, dynamic assistant professor at U.C. Berkeley. "Call me Andy," Professor Lange says, while we walk through his spacious lab with a large assortment of equipment and tools. Several graduate students are there, bent over their desks. Professor Lange jokes with them as he walks by.

Andrew Lange is friendly, collegial, and to the point. He takes me to an adjacent room, where he points to an instrument under a dust cover.

"This is our FIRP," he says, "almost done."

"Who built it?"

"A post doc who left. He didn't finish it."

"Why?

Andrew quickly steers the conversation into a different direction. I don't want to dwell on it either, thinking to myself that something must have gone wrong with that post-doc, and that Andrew would expect me to finish the job. I pull up a chair and sit down to get a good look of the FIRP. Two stainless steel spheres, each the size of a grapefruit, connected with a maze of tubes and mounted in a tight space. Andrew hands me the technical drawings, a whole folder, too much to go through now.

"Who designed it?"

"I did," Andrew answers casually, so as to hide his pride. Then he adds with a smile: "You have to finish it. Money is not a concern. Tell me what you need."

Not so fast, I tell myself. I haven't made up my mind yet, whether I'll accept this position. Somewhere I don't like the idea of taking on a half-finished job. Who knows how well this FIRP has been designed? Who knows how well this piece of hardware has been built? Why did the former post-doc leave in the middle of the project? Can I trust what he has put together? I'll have to look carefully at the design. It's a heat exchange cooler, I've read. I am familiar with the principle. I used one myself at the ETH. The FIRP is designed to work with two reservoirs of activated carbon, high surface area, to differentially absorb helium-3 and helium-4. The principle is clear but working with liquid helium-3 and helium-4 is tricky, to say the least.

Andrew seems to sense my hesitation. He pulls up a chair and sits down next to me. "As I said, money is not a problem. Don't worry about it. I've NASA funding." He then takes the conversation into a different direction, talking about the department and the university, about California and how great it is to be in the Golden State.

"I grew up in Illinois," he says with a big laugh. "What a miserable weather compared to here."

He shows me more of his lab, introduces me to his students, takes me to the head of the department for a formal greeting, and at the end to Professor Paul Richards, an old friend of Jørg Olsen's.

"I heard from Jørg that you might be coming," Professor Richards says. "Jørg calls you one of the world's top experts in liquid and superfluid helium. These are wicked liquids, aren't they?"

After the Berkeley visit, I take BART across the Bay back to San Francisco. Papa waits for me at the Glen Park station. We walk up the hill to the house on Sussex Street, which Papa and Saya recently bought. There we talk until late at night about the Berkeley offer and whether or not I should accept it.

I like the project, though the idea of taking on a half-finished instrument of the complexity of a He-3/He-4 cooler gives me pause. Andrew presented it as if it were a simple task. "Just finish it...money is no problem."

I like the idea that I can stay with Saya and Papa in our San Francisco house, that I can travel to Japan as part of the IRTS project and work there, for months at a time. I like the idea that Andrew Lange is competent, ambitious, and smart. He seems to be extremely interested in having me.

I should say yes. Saya and Papa agree. In the morning I call Andrew Lange and say yes.

In April 1991, I fly to Japan with Andrew for two weeks to meet Professor Matsumoto at the University of Nagoya, who is leading the Japanese group. There will be three more instruments on the IRTS, two being built in Japan, one being built at the NASA Ames Research Center in Moffett Field. But only mine, the FIRP, requires extreme cooling to temperatures below one degree Kelvin.

All the time during this trip Andrew is nice and amiable. He introduces me in glowing terms to our Japanese partners, assuring them that the FIRP will be ready for the launch. At the same time, I learn that the IRTS—like any major satellite project—has to abide by a very strict timetable for the delivery of every single component that goes onto the satellite. And every piece of hardware has to be thoroughly tested.

After our return to Berkeley, my first task is to familiarize myself with the still inoperable FIRP cooler. I spend hours and days pouring over the technical drawings to understand every detail on which its design was based. I see that it was a replica of similar laboratory-scale He-3/He-4 heat exchange coolers, but packaged differently to fit the geometry that would meet the special requirements of the IRTS and the very limited space on the platform. I am trying to understand how the liquid and superfluid helium are supposed to work under microgravity conditions. They are indeed wicked liquids, as Professor Richards had called them, difficult to handle even under Earth gravity conditions in the lab. After thorough examination, I am still not quite convinced that this design will work as specified. Only testing can tell.

Testing takes time. I confer with Andrew about the procedure. Back in Zürich I had done many experiments at liquid helium temperature, using both He-4 and He-3. I learn that the FIRP the previous post-doc built has not yet been tested in its operational state to liquid helium temperature. Nor has any test been done to achieve subKelvin temperatures by putting the liquid He-4 and He-3 to work.

Cooling takes hours. First the entire system has to be evacuated, then comes the first cooling stage to liquid nitrogen temperature at seventy-seven degrees Kelvin above absolute zero, followed by cooling to liquid helium temperature at four degrees Kelvin. Finally, everything looks stable and I can begin with letting He-3 into one of the grapefruit-sized compartments filled with high surface area activated carbon.

Suddenly a very loud snap. The vacuum is instantly lost, and I have to close the shut-off valves as fast as I can. Something must have happened. Something bad. Something ruptured inside the system. I call Andrew. After the whole system warms back up to room temperature, I open the outer enclosure and then the inner one. Yes, a crack between a steel tube welded to the inner steel compartment, which held the liquid He-3. All He-3 lost—some $100,000 worth of He-3. Just lost, blown into the air.

"What's that?" Andrew looks angrily at the broken welding seam. I describe to him in detail how I had cooled the system down, very slowly and step by step, exactly to protocol. I document what I tell him by printing out the record of my cooling procedure, which I saved to the computer.

Andrew calms down after a while. "Stress fracture," he says. He shrugs his shoulder, as if the damage had nothing to do with the design—his design.

I am tempted to say that stress fractures can be avoided. Jørg Olsen was very clear on this in Zürich. He dedicated a major part of his class to designing liquid helium cryosystems in such a way that stress fractures will not happen. When a stress fracture happens during cooling, it's nearly always a design flaw. By studying the technical drawings, I noticed this weakness. I had pointed it out to Andrew, because—after all—this FIRP was his design. He didn't want to hear any of it, got even aggressive when I put my finger on this very spot in the technical drawings. Now, with over $100,000 of He-3 lost, he just shrugs his shoulder and walks out of the lab, mumbling under his breath, "Damn it…stress fracture…."

It will take at least three months to procure the amount of He-3 with the purity we need for our project.

During the waiting period, I return to Nagoya to work with Professor Matsumoto on some other question, namely how to integrate everything into the free-flyer platform. The greatest challenge is the main liquid helium cryostat to cool the beryllium mirror that will focus the starlight onto the FIRP down to four degrees Kelvin.

Together with the Matsumoto team, we calculate the optical path from the mirror to the bolometer array in the focal plane. We have to accurately determine how much the distance between the mirror and bolometer array will shrink during cooling to 2.8 degrees Kelvin, and nevertheless remain in focus with the precision of a thousandth of a millimeter.

Back at Berkeley, I take time to focus on testing the existing FIRP He-3/He-4 system. Andrew stops by almost daily. I keep him informed about every step, but he is showing signs of mounting impatience. I have to tell him that the cryostat is not performing to its full specifications. No problem to reach liquid helium temperature but every attempt to cool the system down into the milliKelvin range has so far failed. I spend twelve hours a day and more in the lab. At night, I sleep on a sofa that I have dragged into my office so that I can be back at work early in the morning.

Reaching 300 milliKelvin is necessary to cool the FIRP bolometer array so that it can register the light emitted from deep space, where the interstellar and intergalactic dust can be as cold as two degrees Kelvin. The cooler the bolometer array, the clearer the images that we'll be able to collect during the IRTS mission.

Whatever I do, however slowly I cool this darn FIRP refrigerator, it just doesn't reach 300 milliKelvin. Not even close.

During one of Andrew's visit to my lab, I finally tell him that there is a flaw in the design of the instrument, a fundamental flaw. I show him my laboratory journal, where I'm keeping a detailed record of every test I do, and what the results are. The closer I look, the clearer it becomes. For some reason, this cryostat doesn't work as it should. Somewhere between the He-3 and He-4 compartments, there must be a thermal short. This short allows heat to flow in a direction it is not supposed to flow. That's why the heat pump doesn't reach the required 300 milliKelvin temperature.

Andrew gets increasingly agitated. "What are you talking about?"

"But it doesn't go down to the milliKelvin range. Barely reaches 2K, only a little bit colder than liquid He-4. It stubbornly refuses to go lower."

Andrew impatiently taps his finger on the benchtop. "My design is not at fault," he repeats. "It's perfect. You just have to complete the instrument."

"But it does not cool down as you can see for yourself."

Suddenly Andrew turns belligerent. "Fix it, just fix it. You are here to fix it." He storms out of the lab.

Only twenty months to the launch date of the IRTS satellites. By now the FIRP should be in working condition, tested for all its basic functions. I should now be able to start the series of engineering tests, submitting the cryostat to the hefty vibration tests that mimic what happens during launch.

I have to redesign the FIRP refrigerator. No other option.

When I mention this to Andrew, he loses his temper. He shouts at me. Behind his anger is fear. A failure of the FIRP would most likely ruin his career. After all, he is considered to be one of the most promising young scientists in astronomy and cosmology in the world.

"Fix it! You are here to fix it!" he shouts at me. "Fix it, tweak it...I tell you, tweak it!"

"If there is a fundamental flaw in the design, you can't fix it by tweaking it, no matter how much and how long you try."

"Oh, come on! There must be some way around it. That's why I hired you. Just go and fix it."

"I must redesign it...from the ground up."

"Not enough time!" Andrew bangs the door behind him. Now what?

I go again through the technical drawings, one by one. I'm looking where there could be a heat leak. I don't find any obvious spot, but the design, how the two compartments are put together, does not look right to me. Surely, somewhere a heat leak develops once the temperature drops below that of liquid He-4. I remember how Jørg Olsen used to say that the weakest point in any subKelvin cooler is the coupling between the two compartments. He had me build one bench-top version for our lab at the ETH and it worked on the first try. My challenge now is to repeat this ETH success for the FIRP.

It becomes obvious to me that Andrew does not really understand how a He-3/He-4 heat exchange cryostat actually works. He underestimates the difficulties when helium becomes superfluid and the heat goes everywhere.

I have to clean up the mess.

Only eighteen months to go until the launch of the IRTS satellite—considered too short to start all over with the design and construction of such a complex, centrally important device. Andrew is in panic and screams at me every day.

I continue unabashed. It's no longer Andrew's project, but mine. I want it to succeed. I work intensely with our wonderful physics machine shop people. They help me rebuild the entire system. They are putting in many extra hours, often late into the night. They are doing a fantastic job, including perfect welding seams. Very smooth. After a few months my new FIRP takes shape. From the outside it looks very much like the old FIRP, but inside are crucial changes.

On the first try I reach 260 milliKelvin—the temperature needed to record the far infrared radiation coming from near the edge of the universe—light that has taken some ten billion years to reach our little corner in the Milky Way galaxy. I repeat the test—again 260 milliKelvin.

My FIRP works.

At NASA Ames, I test the survivability of the system, cryostat plus electronics, by putting them through grueling high vibrational loads mimicking those during satellite launch. I put them through extreme temperature cycles. I test them under high and ultrahigh vacuum conditions.

My FIRP passes all tests.

At the same time, I obtain all technical and bureaucratic approvals required by NASA and the Japanese Space Agency for every piece of hardware that goes onto the IRTS platform and into the satellite. I finish everything on time and send it to Japan.

IRTS launches on schedule from the Japanese space center on Tanegashima Island.

In space, the FIRP functions flawlessly. It keeps its sub-Kelvin temperature during the entire mission—the lowest sustained temperature ever achieved in space for the longest time. It sets a world record.

Andrew Lange never acknowledged my contribution to the success of the IRTS mission. Instead he has taken all credit for himself.

AMELIA

Amelia arrives. I pick her up at the San Francisco Airport.

"Oh, Mino, I'm nervous," she says, with her melodious Swiss intonation and a shy smile. Her ponytail and her voice bring happy memories of Switzerland.

I put my arm around her. "Wonderful that you are here! You haven't changed a bit."

I take her to the hotel where I've made a reservation for her, and we sit in the nearby coffee shop. We talk about what she would like to do and which places to visit.

As we're leaving the coffee shop for dinner, Amelia stops for a moment, and tells me, "You have lost weight. You look taller than I remember."

I might have lost ten pounds or so during the weeks and months I call my FIRP time. "I'll be getting them back now that you are here," I cheerfully reply.

As Amelia and I walk hand in hand through the streets of San Francisco, it feels like we were just leaving my computer room at the ETH. Maybe we are heading for the Berner Oberland to see the gentians. We talk and talk about everything that comes to our minds.

We visit the Science Museum and the Aquarium, then drive across the Golden Gate Bridge to Sausalito. I read to her one of my favorite poems by Lawrence Ferlinghetti, which says: "The light of San Francisco is a sea light" and ends with the powerful words: "and in that vale of light the city drifts anchorless upon the ocean."

We cross the Bay Bridge to Berkeley and I show her the lab where I've been spending so much time during the past two years. I tell her about my FIRP project and what I'm discovering in the endlessly deep sky.

We drive south to the Monterey Bay, where I have been scuba diving a few times. It's an otherworldly feeling to swim below the ocean surface, to know that the sea bottom is one to two thousand meters below, and to see the ever-changing glitter of the sunlight reflecting from the waves above. And there are so many sounds in the ocean: the drawn-out deep calls of whales in the distance; and the faint humming, cracking and clapping, bristling and whistling sounds that the fish, the crabs and shrimps constantly make. I tell her about the great white shark that once glided past me with frightening elegance.

We drive north to Point Reyes National Seashore, where ocean waves are breaking on the rocks with thunderous noise. Point Reyes is the windiest place along the Pacific Coast, a spectacular display of jets of whitewater shooting high up and sending a constant spray across the land. Since it's late

March, the season of the annual Pacific humpback whale migration northward, we are lucky to spot several of them not too far offshore.

Amelia is drawn to the lighthouse, which can be reached by going down more than three hundred steps. Since the sixteenth century there have been innumerable shipwrecks right here, along this shore, because the currents are so strong, the roaring winds are unpredictable, and treacherous rocks hide in the legendary fog. Usually lighthouses are built high up so that they can be seen by ships far out at sea. But in the case of Point Reyes the historic lighthouse was built low to shine the light out to sea below the fog layer.

Leaving the roaring waves behind us, I tell Amelia that the Point Reyes Peninsula sits on the San Andreas Fault and has been moving north-northwest for millions of years. This stretch of land is the remainder of a once large tectonic plate, most of which has been subducted under the North American continent, piling up mountain ranges as far to the east as the Rocky Mountains.

"So, this stretch of land will one day become an island in the Pacific?" Amelia asks.

"Yes, probably in a few million years."

"Let's stay..." Amelia smiles and takes my hand "...stay for a million years."

We find a bed and breakfast place where we can stay. It's going to be our first night together.

With sundown the wind has turned into a gentle breeze. The last shade of red dies away over the horizon. We find a place for a slice of pizza and then take a stroll along the sandy beach of the Drake Estero. The darkness is just enough to see the brightest stars but also the sand on which we walk. The air is getting more chilly. We cuddle together as we return to the bed and breakfast place. We cuddle together all night.

In the morning, after coffee and toast, it's time to drive back to San Francisco. We spend the afternoon with Saya and Papa. In a corner of our living room Amelia spots on the wall the photo of the Pope with Grandpa in the full regalia of a high Shinto priest. Papa stands on one side of the Pope. Next to Grandpa Saya holds me in her arms. "Oooh!" Amelia gasps, making the sign of the cross. "What is this? When was it? Is this you?" She points at me kicking my legs from underneath my baby kimono. I was six months old at that time.

Saya tells her that, in 1962, Grandpa came to Rome to confer with Pope John XXIII about Shinto and whether Shinto should be represented at the upcoming Second Vatican Council.

It was a beautiful warm summer morning, July thirtieth, 1962 to be exact, less than three months before the start of the historic Council. Pope John XXIII had welcomed us in his cozy study in Castel Gandolfo, the papal summer residence on the shore of Lake Albano. The ceiling, walls, floor, and furniture all emanated an exuberant, joyous Rococo impression. Once in a while melodious chirping of birds in the castle garden filtered through the open doors into the room.

The discussion was becoming ever more animated and passionate, how religions in the world could be and should be more open to each other, how they should work together instead of amplifying the differences between them, how they could turn away from greed, envy, and hatred and focus on promoting cooperation. In the Pope's voice there was musical cadence, rhythm, tempo, and beat, his hands often stretching out, sometimes drawing half circles, at other times tapping his chest.

Saya later said that I stayed calm on her lap, keenly observing the Pope, but I apparently felt that, as a six-month-old, I also had a few things to say at this occasion. I must have kicked my legs, uttering sounds like *"Aaah, da-daaah, daaah, aaah."*

Before the audience began, Saya walked down a long corridor, looking for a place to sit down. She found a straight chair without armrests. There she sat in her heavy silk kimono, nursing me. A few monseigneurs strolled by, back and forth, coyly smiling at us. I must have suckled for a long time until finally satisfied. That's when a monseigneur guided us through the sun-filled arcade to the Pope's study.

Now in the middle of the discussion between the Pope and Grandpa I stretched my hands out, saying louder, *"Aaah, daah."* John XXIII took me on his lap and I must have given him a big smile. I touched the Pope's face with my hands.... *"Aaah, daah."*

Holding me on his knees with one hand and gesturing with the other, John XXIII slipped from French to Italian. He spoke about his mother, how she was, how he missed her, how he thought of her whenever he was being carried on a palanquin, his sedia gestatoria, through a throng of people, how he wished to see his mother's face among the many faces surrounding him. It was as if the Pope melted back into his childhood, his eyes tender and affectionate, uttering mama and bambino.

The time was up. John XXIII handed me back to Saya. A photographer came to create an official photo of this wondrous meeting. Two monseigneurs escorted us back, down the long arcade through the garden of Castel Gandolfo and to the Gate, the sun high in the sky. The Swiss Guards saluted us in their bright red, orange and blue uniforms, whose design, according to the legend, was the creation of Michelangelo. Saya says that I still managed to smile at them and wave. Then I must have fallen asleep in her arms with the relief of a memorable mission safely completed.

———

I drive Amelia to the airport for her overnight flight back to Zürich. We still have two hours before she must go through security. We sit in the corner of a coffee shop. I am sad and nervous. I don't want to let her go. She knows that I'll be returning to Japan to start with the analysis of the huge amount of data that the IRTS mission has produced. Thousands, tens of thousands of unique images to be extracted from the digitized FIRP data, images of galaxies far away, stretching almost to the edge of the universe. In addition, there are thousands of images from interstellar regions, where stars are born out of huge dust clouds thousands of light years across.

"How long will you be in Japan?" Amelia asks.

"Hard to guess at this moment—two years, maybe."

"So long...."

"Would you come to visit me in Japan?" I ask, but she doesn't answer.

Instead she says: "Isn't it your turn to come to see me in Switzerland?"

We talk about her plans for the future. She just started her first job as a biology teacher in St. Gallen, her hometown, but she has ambitions to go back to the ETH for a Ph.D. program in biology.

"When?" I ask.

"Perhaps in the fall."

"Then you're also busy for at least two years, maybe three."

"Maybe...."

It's time to go through security. We still stand there as if we could carve off a few extra minutes. Amelia puts her arms tight around me. "Next time you come to Switzerland, promise?"

I promise.

JAPANESE WORKINGS

I've found a one-bedroom apartment in Sagamihara, where ISAS is. ISAS stands for the Japanese government's Institute of Space and Astronautical Science. I'm here to work on the data, billions of data points, that have been produced. I have to sift through these values for cosmic ray hits. The cosmic ray hits have to be removed. Then I'll arrange them so that we can create images of the sky. It will surely take a lot of time.

My one-bedroom apartment is on the second floor of a two-story house with a typical Japanese design—low ceilings, a small living-dining area, a tiny kitchen, a bedroom with a built-in closet for the futon, and a Japanese-style bathroom.

The tatami, which cover the floor in my bedroom, and the fusuma sliding doors, which separate the rooms, are all brand new. I learn that every time a new renter moves in, the owner has to put in new tatami and sliding doors. That's the law, no matter how long the previous occupant lived there and no matter how spotless the place. The costs for this rather extravagant expenditure are included in the amount of money to be paid up front.

That's the way it's done in Japan. So I keep quiet, though it hurts giving away four months' worth of rent. Yes, the Japanese have *shiki-kin*, a security deposit, and *rei-kin*, a "thank-you money." Plus they charge first and last months' rent, which altogether amounts to a heavy up-front expense.

Unlike in America but like in Europe, Japanese apartments are rented without any kitchen appliances and without a washing machine. Luckily for me, however, the Japanese mercilessly discard lots of appliances and furniture, still in perfect shape, simply because they are last year's design. They are up for grabs.

Below my apartment on the first floor, lives a man about forty years old. Every evening a strong scent of grilled fish, curry rice, or something else rises to my window. We greet each other casually, but one day, when I start talking to him, he introduces himself as "Sato," with a bow. He invites me for dinner.

Before leaving for Japan Saya gave me half a suitcase full of small gifts, tokens from Silicon Valley, in case I'm invited for dinner or any other occasion. I give Mr. Sato a booklet of postcards with color photos from California. He confesses his hobby is photography. He looks at the photos with critical eyes and thanks me profusely. He tells me where to buy the freshest fish, the best oysters, and other high-quality catch. "There are three supermarkets around the corner, all competing, but their fish is already one day old," he tells me. "To find really fresh fish, today's catch, you have to go to the fish market—early in the morning. Be there by six A.M." He gives me directions to the nearest fish market. He offers me sake, rice wine, which I politely decline. He pours hot sake into his little ceramic cup and drinks it with one gulp.

After several gulps, Mr. Sato loosens up. "Japanese women are no good," he tells me. "My wife ran away a few years ago—can you imagine—with a man she had just met at a convenience store. Nowadays you often hear the same story. Japanese women are no good."

He refills his sake cup. "Women used to be better in days past. Nowadays they don't want to marry at all or they marry many times. It's chaos."

To commute between my apartment and ISAS I buy a bicycle. The streets are narrow, with narrow sidewalks, and people fill the streets, strolling, jogging, marching, or bowing to each other, holding up traffic. I slowly develop an acrobatic skill to meander on my bike through the crowd. I'm full of admiration for the delivery boys steering their bikes with one hand and balancing on the other a stack of five or six bowls of udon soup.

ISAS is a serious place. We have serious meetings, sitting

around a big table. Usually the meeting begins with an assistant professor or a Ph.D. student making a presentation. Afterwards we keep sitting. The purpose is to discuss important scientific issues, but everybody just sits around, eyes half-closed, arms folded, and heads tilted. Everybody adheres to what appears to be an age-old tradition to read each other's minds. Now and then, the professor at the head of the table groans and clears his throat.

Between the meditation sessions, we go for lunch. At lunch nobody talks about the issues concerning which the meeting has been called. Instead, everybody is busy exchanging last week's football or baseball scores. What the Tigers or Dragons could have done better during their last game. We finish lunch and go back to the meeting room for the next meditation session.

This ritual continues for days or weeks, until everybody feels that a consensus has been reached. I have yet to figure out how Japan works and how the Japanese get things done without the heated discussions we customarily have in Europe and America.

I do tedious data analysis, sit through countless meditation sessions, travel to Shikoku to inspect instruments a company is building for my project, and write for a dozen or more publications. I also enjoy the food wherever I go. I find that food at Denny's, McDonald's, Kentucky Fried Chicken, and Burger King tastes far better than it does in America at the places with the same names. Add Italian and Spanish food, Swiss, French, Greek, Chinese, Korean, Thai, and Indian cuisine. I don't know how the Japanese do it.

In addition, ordinary people I've encountered are remarkably helpful and deeply honest. Once I lost my wallet at the Machida station, where I was changing trains. I reported the loss to a uniformed station officer without expecting to ever see my wallet again. Three days later, it was sent to my ISAS office via registered express mail. About thirty thousand yen in cash, equivalent to three hundred dollars, were untouched. My ID was there, as well as several pictures of Papa and Saya, a small photo of Amelia and me at Point Reyes, my library card, and many receipts. It looked as if the person who had found my wallet never opened it. I wanted to send to the honest finder a thank-you letter with some cash reward, but when I called the station officer, he told me that the finder did not leave a name and contact address.

The Japanese are affluent. They throw away millions of tons of food every day. Neither they nor their economy are dependent on whale and dolphin meat. Not like the Inuits in Alaska, arctic Canada and Greenland, for whom whale meat is and has been essential for survival since time immemorial.

"Why don't the Japanese stop slaughtering dolphins by the tens of thousands every year?"

Most whom I ask don't know that the slaughter is going on, that whole families of dolphins are driven in narrow-entry bays, from where there is no escape. Then they are butchered and speared and clubbed to death by hordes of local fishermen.

I ask but most of those, who confess to have never heard about this bloodbath, nonetheless defiantly retort: "It's an age-old Japanese tradition."

If I persist, they say: "Why not? You in the West have been killing cows, pigs, and chickens for centuries by the millions."

"Yes, that's true. But where in Japan can you find whale or dolphin meat for food?"

Most have never eaten whale or dolphin meat and don't know anybody who has. Yet, they repeat, "It's our tradition. You have no right to criticize our tradition."

I talked to Junichi Tomono in Tokyo, who has written a book about the dolphin slaughter. He confirms that, whereas

whaling has been done for many centuries in Japan's coast-
al waters, dolphins have never been driven into bays, from
where they cannot escape, and killed by the tens of thousands. The so-called "tradition" of slaughtering dolphins was
first mentioned in the 1960s. Why? The best explanation
seems to be that, because the dolphins feed on the fish the
fishermen want to catch and sell, the fishermen view them as
competitors to be eliminated. A pretty bloody application of
Darwin's selection of the fittest.

Strange how hard it is to find anybody in Japan who
can engage in a real discussion. Most whom I met, even in
scientific circles, tend to clam up when confronted with critical questions that beg for critical consideration. They give a
stereotyped reply, often repeating themselves word by word.
Or else they shut up and go away. I realize how different the
Japanese are from many of us in the West. We are trained
to communicate with words, formulate ideas with words,
defend our ideas with words. In Japan, the logic is different. Their thought processes don't work like ours. Ideas are
communicated in much more non-verbal ways, though I still
haven't figured out how.

I've exchanged many letters with Amelia but now we are
switching to email. Her master's thesis about the gentians has
been published and seems to have made a big impression on
many people in Switzerland. Few were aware how bad the
situation had become with the digging out of the gentian
roots for brewing bitter spirits. In many valleys where gentians had always been abundant, few or none are left.

A grass-roots movement has grown, inspired by Amelia's work, to protect the remaining few and re-introduce the
alpine gentians in the wild. She has been asked to lead the
effort. It involves a lot of travel and climbing into remote valleys high up in the Alps. I think of our first climb through that
vertical section of bare rock up to the hanging valley. Amelia

had never worn a harness to secure herself during such short but dangerous transects. I hope she is safe.

"You should come to Switzerland," she pleads in every message. Of course, I want to see her too and promise to come as soon as I'm getting a bit of relief from my tight project deadlines. Until then, I plead back with her, to be careful when she has to reach the high valleys where the gentians bloom. When I close my eyes, I see us, Amelia and me, standing amidst the waves of azure blue gentians blooming on the slopes of the hanging valley. Remembering how we felt—most intoxicated.

I frequently cross the Pacific to fly to Washington D.C. and back, because there is a follow-up U.S.–Japan satellite project in the making—even more ambitious than the IRTS—which is still very much in the planning stage, with no approved funding yet. I stay at the Caltech High Performance Computer Center in Pasadena, where I complete some of the most demanding computation tasks for my FIRP data analysis. Michael Bicay, who was my program manager at NASA Headquarters at the beginning of the FIRP project, happens to be there too, analyzing his astronomical data. It's wonderful to reconnect.

Once I stay for a weekend with Saya and Papa in their new house in Los Altos. I spend hours with Papa, discussing our next science paper. We quarrel and struggle over this or that. He is so slow. Doesn't get it. I'm growing impatient. Makes me mad, but in the end we get the work done. A pretty good paper, I think, about the interstellar dust clouds that obscure so many stars in the Milky Way. A surprisingly large fraction of those nano-sized dust grains is made up of organics, delicate organics floating up there in the vastness of space, surviving the onslaught of very harsh radiation. How is this possible? Papa has an idea. After we finished writing our paper, we send it off to the best astrophysics journal.

In the evenings, the three of us take long strolls in the Rancho San Antonio Park above Los Altos, talking and laughing. When the night sky is clear, we drive up to Skyline Boulevard to watch the stars. One most spectacular sight is Comet Hale-Bopp rising high, showing off its long tail.

In the last month of 1999, after I've spent almost three years in Japan, the NASA Goddard Space Flight Center in Maryland sends me an offer to work there. Almost at the same time I receive an inquiry from the ETH in Zürich, whether I would be interested in applying for a professorship in Astrophysics. The first idea that flashes through my mind is Amelia.

I call her. No answer.

I send her an email. No reply.

I look up other contacts I have for her. She once gave me her parents' phone number. At the fourth or fifth try, her mother picks up.

Amelia is no longer. She died a week ago in a mountaineering accident in the Berner Oberland. Together with two others from the gentian project, she had climbed through a vertical wall to reach a hanging valley, her mother told me. On the way down, an anchor failed, and her rope slipped. Her companions could not prevent the fall. She died on the spot.

After the phone call, I sit in my chair for I don't know how long, gazing at the darkness outside the window. Many images rush through my mind, fragments from the past. No future.

I send an email to my friend in Zürich that I'm not interested in applying for a professorship at the ETH—not anymore. Instead I accept the offer from the NASA Goddard Space Flight Center, just outside Washington D.C.

NASA

PANTA RHEI

At NASA Goddard I plunge full throttle into the new project, designing and building bolometer arrays for NASA's SOFIA aircraft, the world's most advanced airborne astronomical observatory. It will be mounted in a Boeing Jumbo Jet, modified to fly very high, above ninety-nine percent of Earth's atmosphere. At this altitude, using its German-built telescope with a mirror, two and a half meters across, SOFIA will be able to record the light from almost the same distant sources as FIRP has done. The bolometer arrays for SOFIA are going to be cooled with liquid helium, but there is no plan to go down to subKelvin temperatures.

I realize that through FIRP, I've entered the esoteric field of cosmology, looking at the beginning of the universe and how it evolves through time and space. There are many generations of stars, and the question still looms large how time itself changes with the expansion of the universe.

Some cosmologists say that time and space are created, but this creation is of a different kind than some of religious authority proclaim. It's not a once-in-time creation in the past but the continuous process of something changing all the time, something being created, over eons, maybe even for eternity. A faint reflection of this idea can already be found in the writings of Heraclitus, the pre-Socratic Greek naturalist and philosopher who coined the phrase *panta rhei*—everything flows, everything changes—like the water in a river, the trees in the forest, the mountains on the horizon. Everything is continuously created and recreated in time and space.

Nothing stays. I think of Amelia and sadness overwhelms me again.

After having spent almost a decade in the more contemplating mood of cosmology, where the dimensions are measured in light years, I return to the other world, that of tiny dimensions, where the scale of things is measured in micrometers and nanometers.

NASA's Goddard Space Flight Center is a powerhouse. It has a state-of-the-art Detector Development Laboratory, with all facilities to micromachine silicon chips that we need to build the bolometer arrays. It has an advanced optical characterization laboratory and a superconducting electronics lab with very sophisticated software.

In addition, there is a standing collaboration with the University of Maryland to take advantage of its ion beam milling facility, as well as with NIST, the National Institute of Standards and Technology, which also has unmatched facilities to which I would have access, when the need arises.

While in Japan, I told several contacts at NASA Headquarters in Washington, D.C. I would be coming to Goddard soon. Within days I received an invitation to the science debriefs that are given every month to the top brass.

After two and a half years, however, I am now happy to leave Goddard. The basic design of the bolometer arrays is done. The first prototypes have been built and tested. Improved versions are on the drawing board. However, the SOFIA aircraft, where our bolometer arrays are to be installed into the heart of the fine telescope with its two-and-a-half-meter-wide mirror, is experiencing delays after delays, mostly because of cost overruns. The original plan was to have a large door in the front part of the jumbo fuselage, which can be opened and closed while the plane is flying at full speed at high altitude. This design had to be abandoned or rather modified. Now they are putting in a door closer to the tail section. The wings of the airplane also had to be modified to allow it to reach much higher altitudes than passenger jets. Delays after delays.

The bolometer group is treading water. I keep suggesting several alternatives to take advantage of this lull in our project, but when working for the government, innovative ideas are not always welcome.

The Air Force Research Laboratory in Dayton, Ohio invites me to join their Sensor Directorate. I visit the place and find it interesting. So, I accept the offer.

Soon after starting at the Air Force Research Laboratory, I meet Barbara Wilson, whom I knew from CalTech. We talk about nanotechnology, which is not yet really developed at the AFRL. This makes me think again about my time at the ETH, when I built from the ground up the first Scanning Tunneling Microscope, after Rohrer and Benning had shown at IBM that this principle works. By now there are dozens of variants of the same idea, most notably the Atomic Force Microscope, but I still remember the sensation of pride and joy rushing through my body when the first images of individual atoms appeared on my computer screen, the expected hexagonal pattern of carbon atoms in a single crystal graphite sheet. When I saw those atoms, I knew that in a few years' time we would be able to routinely view, measure, and manipulate materials down to the dimension of a few atoms, truly nanometers in size.

After Barbara Wilson leaves AFRL, I continue as the Chair of the NanoScience & Technology Working Group, defining and implementing over the course of the following year the first nanotechnology strategic plan for the entire Air Force. As part of this activity I've secured for AFRL substantial new funding.

MINOSATS

Now I am at the NASA Ames Research Center in California. It's strange how it all happened and how quickly too. I was visiting Saya and Papa in Los Altos. Papa told me that he had an appointment in the morning with Michael Bicay, the Director of Science at Ames.

"I know Mike," I said instantly.

"How?"

"He was Program Manager of NASA Headquarters for the IRTS mission. He's known me since my Berkeley years. Later, when I used the supercomputer at Cal Tech, Mike was there too."

"Why don't you go to see him instead of me?" Papa suggested off the cuff. "Just as a surprise?"

Yes, it was a surprise. Mike had heard that I had done the nanotechnology portfolio for the entire Air Force, while I was at the AFRL. He told me that NASA Ames was looking right now for someone to lead its own nanotechnology effort. He asked me whether I would be interested and sent me upstairs to see Steve Zornetzer, the Deputy Center Director. Steve sent me to see Chris Christensen, the Acting Center Director. One hour later I was hired.

I joined NASA Ames in April 2006 as the Director for Nanotechnology and Advanced Space Materials. A few weeks later, a new Center Director came onboard, Pete Worden.

The NASA Administrator, Mike Griffin, the top gun at the Agency in Washington, D.C., was famous for saying "Smaller, faster, cheaper." A competent engineer, he had grown up in the culture of multi-billion dollar space missions, which included large, heavy satellites and powerful rockets— all quite exquisite but also very expensive.

I liked Griffin's "Smaller, faster, cheaper." I wanted to turn his slogan into reality by weaning NASA off the idea that all space missions must be large and expensive. I met Robert

Twiggs, Professor at the Stanford School of Aerospace and Engineering. He was designing a nanosatellite only ten by ten by ten centimeters in size and was looking for ways to make such a tiny box useful for space applications.

Instantly I saw the huge potential of this tiny box idea and took it to NASA Ames. Pete Worden liked it. Al Weston, our NASA Ames Director of Programs and Projects, liked it. I started with my team to develop advanced concepts how we might fit the many requirements that satellites have to meet into such small boxes and have them function safely in the harsh radiation environment of space.

We worked on it for months. My familiarity with nano-technology helped me enormously. We were able to show that all essential mission requirements could fit into this ten-centimeter cube standard. Minimum energy require-ments. We reduced the size of all components. We did it... we did it.

Well, Pete Worden apparently had not expected other-wise. He put out a NASA Center-wide announcement, in which he listed my broad background in solid state physics, surface physics, laser physics, observational cosmology, infra-red astrophysics, space cryogenics, satellite integration and testing, large-scale data analysis and data mining software, photo emission from low dimensional systems, and nanowire device applications—what a mouthful—*and* he coined the word "MinoSats," which made me feel really proud.

With Pete Worden's support, I massively promoted the NanoSat idea at NASA Headquarters, which was a very hard sell. Some people, when they saw me walking down the hall-way, sniggered: "There he goes...Mister MinoSats."

I travelled to D.C., often alone, and sometimes with Al Weston, to knock on many doors, such as the National Reconnaissance Office, DARPA, the Air Force, the Army, and of course NASA HQ. Some groups at these government agencies marveled at the MinoSats idea and began to implement our vision. I flew to London with Al to start spreading the word of MinoSats across the world and talk about the technical feasibility of competent satellite missions using the ten-centimeter cubesat format. "Yes," we assured them, "we can shrink all components in size, lower the electrical power requirements, and reduce the costs significantly."

However, Mike Griffin, the top gun at NASA, didn't like it. In his mind we were going too far, despite his catchy phrase "Smaller, faster, cheaper." Pete Worden, by now my supporter and friend, stayed firm, even though he then became the target of political backstabbing by members of the House and the Senate from those states where other NASA Centers were located. Those politicians started to question why the NASA Ames Center in California would be allowed to enter into the competitive world of satellite missions. They were afraid that if the idea of "MinoSats" took off, the NASA Centers in their states might lose lucrative contracts for building and launching the traditional billion-dollar satellite projects.

Griffin abruptly cut much of the funding for my nanotechnology efforts at NASA Ames, creating a nightmarish situation. I had a large team, very good scientists and engineers, highly interdisciplinary, cutting across disciplines from physics to engineering all the way to space medicine. They were all dedicated and looking forward to an exciting time with new nano-ideas and new nano-initiatives. Mike Griffin destroyed it all.

Undaunted by the controversies popping up all around, I multiplied my efforts to convince the experts at NASA Headquarters in D.C. Looking back at my travel records, I realized that, in one year, I had spent two hundred days flying across the country, visiting D.C. and other NASA centers like Goddard in Maryland, Kennedy Space Flight Center in Florida, Marshall Space Flight Center in Alabama, Johnson Space

Flight Center in Texas, JPL in southern California—all in an effort to garner support for our new nano-concept in satellite technology.

I'm really convinced a large portion of future NASA Space activities can be and will be carried out by the small ten-centimeter cube boxes. If we run out of room in one box, we can string three of them together to make a three-unit nanosat or six into a six-unit nanosat or even twelve into a twelve-unit nanosat, to accommodate additional instruments and components for complex satellite missions.

The simple reason is that the high costs of conventional space systems can't be sustained forever. As my friend Al Weston points out, in the future MinoSats will revolutionize space technology and amplify the amount of science that can be carried out in space and from space. Thousands of students around the world will have opportunities to expand their horizon and learn new skills with MinoSats.

To handle the massive amounts of data that stream from each satellite mission, NASA Ames is lucky to have the World-Wind program office, headed by Patrick Hogan. World Wind is "Google Earth" on steroids, independently developed by NASA, superbly suited to manage—in real time—huge and complex static and dynamic data sets, adapting on the fly to ever-changing conditions. I've discussed with Patrick how we might join forces to use WorldWind's remarkable features for many of my projects.

However, facing ever shrinking budgets by NASA Head-quarters, I struggle to maintain funding for my team. Pete Worden does his best to help but his budget is also impacted by the decisions at the top.

In order to help my budget situation, I take on the responsibility to be the technical monitor for forty-seven NASA contracts with industry and universities across the country. Huge extra load—and all this without secretarial help. Tomoko Ishihara, an engineer in my group, is kindly taking some of the load off my shoulders, and she is highly efficient and competent. Harry Partridge, my Deputy, sees many projects through that need different levels of attention. His never-ending cheerfulness makes difficult tasks look easy. When the sailing gets tough, I walk over to Building 200, the "Head Shed," and sit in Karen Bradford's office for a little while. Karen is Pete's Chief of Staff. She has a wonderful voice, with a velvet pitch and a soothing timbre.

At the same time, I have been applying nanotechnology to a new neuroscience project, which may open new opportunities for the treatment of brain cancers. Take nanoparticles of iron oxides, a yellow-brownish powder where the individual grains are only a few tens of atomic diameter in size. Tiny, tiny. They are magnetic, more precisely paramagnetic, in fact superparamagnetic, a property only found in nanomaterials where the grain sizes are in the nano-range. These tiny iron oxide particles can be loaded with drugs that fight cancers. By injecting them into the bloodstream and let them be swept into the brain, magnetic fields can be used to concentrate them where they would be needed. There—we hope—the nanoparticles will be able to cross the blood brain barrier and deliver their medically active drug load. All without surgery that always damages healthy brain cells and nerves.

It is great to be at the NASA Ames Research Center. Here I can do things that go far beyond NASA's specific aerospace agenda. I can do things that bring together the knowledge of very different disciplines, capitalizing on what I have learned in the past twenty-five years, applying space technology back to Earth and use it to benefit people down here. I'm really fortunate to work in such a stimulating environment—even though the perpetual worries about funding plunder my sleep.

FIGHTING

GBM—GRADE IV

Late August I was to meet a researcher at the University of California Berkeley to discuss a joint project in neuroscience. Specifically, I planned to talk with him about the possibility of applying nanoparticles to the treatment of brain tumors. There are more than one hundred different types of brain tumors, some benign, others malignant.

However, I didn't feel well. I had a severe headache and felt nauseated. Saya had been suffering from the flu for some time, also with a persistent headache and fever. We both decided that I had caught it from her. She told me to stay in bed and brought a cup of herbal tea. Nevertheless, the headache was wrecking me. I took one aspirin and tried to sleep it off. Didn't work. The splitting headache remained. A second aspirin made me feel slightly better, but now my left hand felt numb. The left side of my mouth was drooping. We wondered what a strange flu I seemed to have caught.

"Maybe a pinched nerve," I suggested to Saya. "Should I go to see our acupuncturist?"

She urged me to go to the Emergency Room. With my horrific headache I drove to the Palo Alto Medical Foundation and checked into their ER.

The attending doctor gave me a strong painkiller. He ordered a blood test and set me up for an MRI scan. The

MRI was relaxing, soothing. It felt like floating in a Jacuzzi.
The technician could not have been better. After the MRI the
attending doctor put me immediately on intravenous dexa-
methasone decadron, a powerful steroid. He explained that
the decadron was to relieve the edema, which was giving
me the headache. He prescribed decadron pills to be taken
every four hours. "Come back tomorrow morning for more
details."

Next morning, I was there together with my father. Dr.
Paul Jackson brought up on his computer screen the MRI
image of my brain. He pointed to a dark blob in my right
parietal lobe, the size of a lemon. He pronounced it to be a
tumor, most likely a glioblastoma multiforme, GBM, grade
IV.

"You are a scientist," he said in a detached tone. "I can
tell you the truth. This is a GBM. There is no cure for GBM,"
Dr. Jackson continued without pause. "You'll live for a few
months."

I was dazed. I looked at the doctor who seemed void of
human emotion.

"What do you mean that I'll live for a few months?"

"If you undergo immediate surgery, followed by six
weeks of radiation and by chemotherapy, I can add a few
months. The best you can hope for is to live for about fourteen months—from today on."

Somehow, I couldn't grasp the fact that he was talking
about me, about my brain, about my life. Everything was
eerie, too abstract to sink in, drifting through space, defying
imagination.

I have been healthy and active all my life, in Europe,
Japan, and America. Except for the occasional cold or flu,
I've never had any illness, surely no serious illness. Yes, a
stiff neck from sitting for too many hours in front of the
computer. I then went to an acupuncturist or did yoga. I did

tai chi and chi gong in Maryland while at the NASA Goddard Center. My work was intense, but in Ohio, during my time as the Chair of the Nanotechnology Group at the Air Force Research Lab, I often drove to the Glen Helen Park, a refuge in an expanse of forest, with pristine creeks and one very old oak tree that I befriended. Now in California, at the NASA Ames Research Center, I spent many weekends among the giant redwoods in Big Basin State Park or on the windy beaches of Half Moon Bay.

I never smoked, never used drugs. I didn't drink except for a glass of wine or champagne now and then. My diet has been healthy with the best green tea, gyokuro, lots of fish, other seafood, and vegetables.

My height has been five feet eleven inches, my weight one hundred fifty pounds throughout my adult years.

A lemon-size tumor in my brain?

How is it possible?

Since when has it been growing?

Why did I not notice it? Why didn't I have any symptoms until now?

Fragments of thoughts raced through my mind.

And surgery? A surgery of my brain. I looked at Dr. Jackson and asked, "How is a surgery done?" thinking about many stories I used to hear from Professor Akert, neuroscientist and Director of Neurology at the University of Zürich, in whose house I lived as a student. In fact, Professor Akert gave me my first job over the weekend to map the visual cortex of spider monkeys. Through my contacts with him I learned what neurologists do, though at that time, in the early 1980s, clinical MRI was still in its infancy and brain surgeries were more or less guesswork.

"Well," said Dr. Jackson in an uppish tone, "we cut your skull open with a saw—in your case on the right side—from the top to the ear, a hole of about 4 to 6 inches. We take out the tumor as well as we can, close the skull with a titanium plate. It will be put under the skin of your skull and hang there."

"Why don't you put the bone back?"

"Takes too long to heal."

"Why too long?"

"You live only for a few months. Easier with a titanium plate. Go home and bring your affairs in order. I'll schedule you for surgery tomorrow morning. Come back at seven o'clock."

On our way to the parking lot, Papa looked suddenly old. He uttered words like "Poor Mino. Oh, Mino. What can we do...? Oh, poor Mino."

This made me furious. I saw in Papa's eyes that he had surrendered to Dr. Jackson's verdict. This made me furious from the bottom of my heart.

I am not surrendering. How can this doctor dare to determine how many months I'll live from today on? A few months—several months—utmost fourteen months? To hell with this seven A.M. surgery. He thinks he can push me around, order me to show up at seven A.M. No way. If I have such a bad brain tumor, I'm going to do research. I'll talk to the best neurologists. I'll determine by myself what to do, where to go, and when to begin. I'm going to fight it, fight it, fight it, fight it, *fight it*.

WHAT IS CANCER?

I barely remember how the next days went by. Many hours at NASA, meeting with our Center Director Pete Worden and other friends, lots of support from everyone, offers from many, who want to donate their accrued sick leave days so that I can fight my tumor without worrying about time off from work. One brings self-cultivated mushrooms that should kill cancer cells. Another tells me to take a spoonful of turmeric daily. I should also take lots of vitamin D. I should not eat carbs but live off proteins. I hear Karen Bradford's deep soothing voice that I'll be all right. Steve Zornetzer signs my medical leave papers and NASA Headquarters approves them in record time.

Deb Feng organizes a party in her house in Palo Alto. Many of my NASA friends come, standing room only. I give a defiant speech that I'll fight it, fight it, fight it. At the end Pete Worden wishes me Godspeed.

At home, I'm spending hours on the web searching for the best option to fight GBM. I learn that GBM can strike patients of all ages from small children onward, even babies. GBM tumors grow fast, very fast. The median survival offered by the medical authorities—surgery, radiation, and chemotherapy—adds only two and a half months to the life expectancy. In Western medicine the combination of surgery, radiation, and chemotherapy is called the Gold Standard. Beyond this Gold Standard Western medicine has little to offer.

I start looking into alternative methods, which many call "integrative medicine." There are about four hundred protocols to go through and to evaluate. Many draw on traditional Chinese medicine and the quintessential Chinese concept of *chi*, Life Energy. This is familiar territory for me—tai chi, chi gong, meditation, herbal teas, acupuncture, and many other forms of helping the body's energies to prosper and to circulate more freely in the body. This also includes massage, diet, and the Indian practice of yoga.

The problem I'm facing is time. All forms of integrative medicine are known to work slowly. They take months, even years to show beneficial results. No quick fixes. But time is what I don't have. I am in a race against my aggressive GBM tumor.

Why does a tumor form in the first place? Cells in our bodies are damaged every day, every minute, every second. They are damaged by cosmic rays, by the radioactive decay of radon in the air we breathe. Every moment cells are hit, damaged or outright destroyed. Everywhere on Earth and all the time.

To deal with this relentless assault, Nature has given us an immune system that has evolved to recognize damaged cells or malfunctioning cells or dead cells. Our immune system has learned to grab them and dispose of them.

However, occasionally the damage may not be distinct enough for our immune system to recognize. The most frightening mutations are those when cells find a way to revert to that very primitive early stage of life that existed on Earth long before there was oxygen in Earth's atmosphere, long before multicellular organisms had evolved, at the time of primitive single cell life, growing in leaps and bounds, forming colonies and clumps, feeding on plain glucose, the simplest of all sugars.

If healthy cells in our bodies mutate to this kind of archaic cells, they can survive and proliferate, but they no longer adhere to the master plan of our bodies.

This is cancer. Cancer cells are our healthy body cells gone awry. Cells that have started to multiply in an uncontrolled way, destroying surrounding healthy tissue, oblivious of structure and genetic function.

Cancer cells have reverted to the lifestyle of those primitive single-cell organisms that dominated the Earth at the dawn of life, billions of years ago, voraciously consuming glucose and glucose only. If perfectly good cells in our bodies revert to this archaic lifestyle, they have the power to increase in numbers, as our body's immune cells stand by, unable to fully recognize the danger, hesitant to attack the multiplying tumor cells.

Once a tumor is established—as in my case, GBM—the odds are poor to defeat it, to eliminate it, and to remove it from the brain.

Within five days after the diagnosis of my brain tumor, I decided to fly with just a backpack to Pittsburgh, because friends at the Carnegie Mellon University campus at the NASA Ames Research Park have told me great things about the Brain Cancer Research Center of the University of Pittsburgh Medical School. In addition, I wanted to fly to New York to meet a practitioner in Traditional Chinese Medicine, TCM, who has supposedly been successful in getting rid of tumors.

WITH GUSTO

I arrived in Pittsburgh. The weather was just grand, reminding me of Zürich. Very uplifting. I looked forward to my meeting with Dr. Daniel Wecht, the neurosurgeon whom Jared Cohn, the president of Carnegie-Mellon University, had recommended.

On the evening of my arrival in Pittsburgh, I forgot to take my prescribed dose of decadron to control the edema in my brain. A few hours after I had fallen asleep, I woke up in cold sweats. Decadron is about thirty times more powerful than hydrocortisone. The ER doctor back in California had warned me that, if I were to interrupt my daily dose of six times four milligrams by more than ten hours, I might die. I was careless, because the half-life of decadron in our body is so short. I can't risk forgetting it again.

I also have to take eleven other medications and supplements. My iPhone is my best friend to remind me of my schedules. Getting through airport security with four pounds of liquid medicines also takes time and patience, as I found out at San Francisco Airport.

The meeting with Dr. Daniel Wecht restored my faith in the medical community. There was so much willingness on his side to listen to me, to understand my needs, and to answer my questions. He never pretended to know everything. He never took refuge in statistics to judge my case. Capable and compassionate, he supported my decision to seek other options besides or beyond Western medicine. He gave me the peace of mind that had already started to slip away. He agreed to supervise my decadron doses and to prescribe MRIs when needed.

The other good news is that a whole-body CAT scan that Dr. Wecht had ordered showed no sign of a cancer anywhere else in my body. Nothing to see. Nothing to worry about. So, my brain tumor is a primary tumor. I have only one enemy to fight—my GBM.

Leaving behind the Pittsburgh integrated neurosurgery team, which surely is one of the best in the country, I am flying to New York. Next comes unchartered territory of my journey. I'm going to see the TCM practitioner, Frank Huo, who claims to have cured cancers, sending them into remission.

"What kinds of cancer?" I ask.

"Liver cancer and pancreatic cancer," he answers after a moment of hesitation.

Hard to believe, but I want to believe. I know from my own experience what traditional Chinese methods can do when applied by a masterful hand. Years ago, while in Japan, I immersed myself into chi gong, the Chinese tradition to gain inner peace by controlling breathing. There are innumerable variations of how to help your body using breathing techniques. Those who master the art can even control their heartbeat. When I came to NASA Goddard in Maryland, I was introduced to Master Ma in nearby College Park. Through him I achieved a level of concentration so deep that I could lower my heartbeat at will. The lowest I reached was fifteen per minute, almost as low as a hibernating bear. Each time I came out of such a session, my body was filled with unimaginable energy that seemed to be radiating from within.

Frank Huo is a man of few words. He listens quietly while I describe my symptoms. I demonstrate how inept my left hand is holding onto anything; a cup, a glass, a pen—everything slips out of my fingers and I don't even notice it. I try to stand on one foot. Yes, it works with my right foot, but as soon as I switch to my left foot, I lose balance.

This is understandable, since my tumor is in the right parietal lobe of my brain. It affects motor nerves going to the left side of my body, to my left hand and my left foot and to the left corner of my mouth, the one that droops.

Frank brings his acupuncture needles and asks me to lie down. He works on me, inserting needles into acupuncture points along the meridians that go to my left hand and left foot. His treatment is soothing, and I fall asleep.

This was Saturday afternoon. Sunday afternoon a second session. Early Monday a third.

Now it's Monday afternoon, and I can type freely again, with both hands, at full speed with hundred-percent accuracy. Without opening my eyes, I can recognize textures and shapes with the fingers of my left hand. I can feel fabric, wood, plastic, or metal. I can feel whether a point or a cutting edge is sharp or dull. I can pull things out of my pocket, and the temperature perception with my left and my right hand is identical again, both for hot and cold. Nothing drops from the grasp of my left hand. I try a cup, a glass, a pen, a piece of paper, and my wallet. Touching my nose with closed eyes works well, too. I can even recognize numbers written into the palm of my hand. No difference between left and right. I can stand on one leg, left or right, with the other folded over my knee.

On my way back to the hotel, a good half-hour stroll by foot, I try to walk briskly. No problem. I try to run. No problem. By the time I arrive at the hotel I'm exhausted with joy.

I can't fall asleep, however. How come the symptoms

go away so quickly and so completely. After all, I do have a tumor in my brain and this tumor upsets the nerve communication between the brain and my limbs. If these perturbations are gone now, after just a three-day acupuncture treatment, has this really changed my brain tumor?

Or, alternatively, does this mean that the GBM tumor is still there as before, but the three-day TCM treatment was able to restore the functionality of the surrounding brain tissue, so that the nerves could send and receive signals as before? I remain skeptical, but it's very interesting nonetheless.

There is much chitchat on the web about spontaneous remissions of cancers, and some TCM practitioners insist that cancers they have treated all disappeared. Yes, it's possible from everywhere else in the body, but not from the brain.

I don't believe that my GBM tumor can disappear. Not so fast. After all, when it was diagnosed two weeks ago on that horrible day in Palo Alto, it had already grown to the size of a lemon, four point three centimeters across, a clump of billions and billions of cancer cells in my brain. Most of these cells are dead because the cancer cells on the surface of this clump grow so fast that the ones inside the clump are starved and die. Spontaneous remission means that cancer truly disappears. However, where can the dead GBM cells go? How can the brain transport them away?

There is the blood brain barrier, BBB, a wonderful protection, which Nature has designed to prevent outside enemies, in particular bacteria, from entering the brain. Because of the BBB, bacteria can't get in and infect the brain. But at the same time the BBB makes it difficult, if not impossible, to remove any unwanted debris from the brain.

I must ask Dr. Wecht to prescribe a new MRI scan, so that I can see what's going on with my tumor and whether it has shrunk.

I finally fall asleep.

In the morning, in the breakfast room, I sit down at the piano in the far corner and start playing with gusto.

Next morning, however, my left hand starts to feel numb again. Things are slipping out of my fingers, and I don't even notice it. This damn tumor.

I have to fight it, fight it, fight it.

I don't know what to think of Frank Huo's treatment. It's amazing how he was able to make the symptoms go away. After just a few acupuncture sessions, I could use my left hand. I felt stable on my left foot. I could walk and walk briskly.

But now it's coming back, this stupid numbness.

Dr. Wecht did not prescribe a new MRI, arguing that it makes no sense to have another MRI after the one in Pittsburgh, barely two weeks ago. However, I still want to know what is happening in my brain and whether my tumor has at least stopped growing.

I find a commercial place where they do MRIs. I go there and lie in the MRI scanner. The result is disappointing. No clear image. The contrast is poor. I can't make out whether the dark lump in my right parietal lobe is smaller or larger than it was in the Pittsburgh MRI.

Frank says that because of his acupuncture and my meditation, my tumor has shrunk. He insists he can feel the tumor when he passes his hands over my skull. I can't believe it, but deep inside me a voice whispers that he might be right. What if my GBM is indeed shrinking?

As long as I'm taking my decadron doses on time, I have no headaches, but I must ask myself whether I'm not wasting my time.

Every night I call California and chat with Saya. She always asks whether she should come, but I tell her that I'm doing fine. I skype with Papa and get regular updates on what's going on at NASA Ames. After I hang up, the dark hours drag on and on. It's hard to fall asleep. Sometime in the early morning hours I must have gotten up. Still dark outside. I fill the tub with comfortably hot water. I take a bath. After having soaked for some time, I step out of the bathtub.

I don't remember how it happened, but I must have slipped on the wet bathroom tiles.

I don't know how long I have been lying there, but after a while I could grab the towel and dry myself. Somehow I made it back to my bed.

I call 911.

I call Saya.

SAYA'S STORY

Mino's screaming voice came through the phone. In the middle of the night, at 2:30 A.M. California time. "I fell in the bathroom. I hit my head!"

"Minochan, Minochan, lie down...please lie down," I forced myself to stay calm. "I'm coming...I'm getting on the next flight to New York."

"911—they are already on their way."

Friedemann is frantically searching for a seat on the next possible flight to New York out of San Jose or San Francisco.

I start to throw things into my travel bag. The phone rings again. This time it's the 911 driver. "His vitals are okay. We're taking him to the Queens Hospital." He hangs up.

His vitals are okay, are okay, are okay.

Friedemann cannot find any available seat on any non-stop flight to either JFK or La Guardia in New York. Nor to Newark. Not even First Class. I have to take a flight via Texas.

In the plane, dozing in my seat, the last weeks come back, a parade of turbulent, often contradictory feelings. Mino has called us every night to tell us about his day. He effused hope that, with Frank Huo's TCM treatment, he was on a promising path. Regaining the full use of his left hand and ability to walk, and even run, filled him with seemingly endless joy. I shared his joy but there was always some lingering fear that the good news was not going to last.

Frank Huo insisted that the tumor was shrinking.

Understandably, Mino wanted this to be true. With every phone call, it became ever more apparent that he was being pulled deeper and deeper into Huo's world of wishful thinking that the power of *chi*, the universal life energy, is capable of overcoming a brain tumor even as evil as GBM.

Why did Mino fall down in the bathroom? Not just the slippery floor. He must have lost balance because of the paralysis spreading through the left side of his body.

Twelve hours later, after a long stopover somewhere in Texas, I arrive in La Guardia.

Now what?

Mino might still be in the Queens Hospital or somewhere else or back in his room in the hotel, the Lexington.

I call Friedemann but he doesn't pick up.

I call Frank, but he doesn't pick up.

My cell phone goes dead.

I don't know the address of the hotel. I only know it's not far from La Guardia.

I'm digging through my travel bag. I have print-outs of all emails Mino has sent us from his hotel, but no hotel address. I find my reading glasses, my credit card, my California ID, $60 cash, somehow two toothbrushes, one tea bag and my old, bulky address book with all Japanese, American, German, Swiss, French, and British addresses that I have collected over decades, but no address for the hotel.

I get into a taxi. The driver starts driving down the street. "You know the Lexington, the Lexington Hotel?"

"No." He is driving full speed.

"It's close to the airport."

"Ain't know nothing." He is jiggering with his whole body to the rhythm of the music blaring from the radio.

"Stop at the next hotel, please."

"Huh? Hotel? Which hotel?"

"That one over there...it's a hotel."

"Huh?"

"Here. Stop. Wait."

"Huh?"

I run into the hotel lobby, and they help me with the address of the Lexington. I give the taxi driver the address and he sets out driving again what seems to be an endless cruising around the same city block.

Finally, here it is, the Lexington.

"My son is staying in your hotel," I tell the reception-ist, young and blond. "Here is his name. What is his room number?"

She smiles. "Room 1811."

I rush to the elevator, up the eighteenth floor. I knock at the door of Room 1811. No answer. I knock harder. No answer. I knock again, calling "Minochan, Minochan."

I jump back into the elevator and drop down to the recep-tionist. "My son doesn't open the door...I'm his mother, just arrived from California. Give me a second key."

"Sorry. I can't give you a second key."

"But I'm his mother."

"Sorry, I'm not allowed to give you a key. Your son must first call us and give me a permission to give you the key."

"But he may be unconscious, he may be in a coma after his fall this morning. It's serious—911 came this morning."

"I know," the receptionist says with her frozen smile. "They brought him back, too. But I'm not allowed to give you the key."

"I'm his mother. He is severely ill."

"Sorry, I can't help you, unless he calls me and gives me permission."

"He has a brain tumor and hit his head this morning. 911 came."

"I know."

"So, give me the key."

The woman continues her smile. "Our hotel policy is never to give a second key to anybody, unless the occupant calls."

"But don't you see? My son may be in a coma. How can he call, when he is in a coma? This is an emergency."

"Sorry, hotel policy is hotel policy. I didn't make the rules and I can't change them." The receptionist never stops smiling.

"I want to talk to the manager."

"Sorry, he isn't here."

"Anybody else?"

Out comes another woman, this time fat and middle-aged.

"My son has been staying in your hotel for three weeks. What kind of treatment is this?"

"When 911 brought your son back this afternoon, he asked us to give the second key to a Chinese guy," said the fat lady. "You should ask that guy to let you in."

Out of courtesy she lets me use the hotel phone to call Frank Huo.

No answer. I call his number again and again. Finally, he picks up. Very difficult to understand him but I tell him, where I am. Almost one hour later he arrives while I'm sitting in the lobby with my back to the ever-blaring TV screen high up on the wall.

"Mino is doing okay...He is fine," Frank says. "Queens Hospital didn't find any bleeding in his head."

After Frank opens the door to Mino's room, I tiptoe in.

Mino is stretched out in his bed, pale, almost gray, breathing lightly. His stubby beard makes him look aged. The left side of his face is distorted and the left corner of his mouth is drooping. He opens his eyes and tries to smile.

I put my arms around him.

"Aaaa, Sayachan," he murmurs.

"Minochan, Minochan, I'm here. Friedemann comes tomorrow. We are both going to be with you." I feel his body relax. He slides back into slumber, a tear in the corner of his eye.

Ever since we brought him to San Francisco Airport, Mino kept telling us not to worry. "I can take care of myself."

Two weeks ago, on the phone, he was elated telling us about his astounding improvements after the acupuncture sessions.

One week ago, Mino told us that his tumor was shrinking. "Only two centimeters now," he said.

"How do you know?"

"Frank says so. He can feel the tumor by placing his hands on my head."

"But your last MRI?"

"Poor quality. I don't know for sure, but it's shrinking, I believe. Frank says, if my acupuncturist in California were any good, he should have discovered my tumor, when it was still very small."

If Frank Huo is so knowledgeable, how come that he can't make the connection between Mino's drooping face and the tumor in his brain. Impossible that the tumor did shrink to two centimeters.

Over the course of the past week I noticed a change of tone during our daily telephone calls. Mino also started mentioning "karma," as if it had something to do with GBM. According to Frank Huo, the GBM comes from his karma. By cleansing his karma, Frank claims to have already reduced Mino's GBM to half its size.

Mino knows that the Sanskrit word "karma," meaning "action," appears already in the sixth century B.C. in
India, before Buddha. It means the actions you did in your
former lives follow you forever through reincarnation. A
never-ending cycle of rebirth.

The concept of karma forms one of the fundamental pil-
lars of Indian thought. Karma sneaked into Buddhism after
Buddha's death in order to explain to the followers why many
people suffer from diseases, accidents, or natural disasters.
Buddha himself never mentioned karma. He denounced the
idea that ill deeds committed in your former life would come
to haunt you in this life. When asked by his disciples about
karma, Buddha famously remained silent. His silence was
so absolute that those around him understood his refusal
to speculate. He was surely aware such speculations could
be turned into a tool to control the minds of the people, a
powerful weapon.

And what about the Christian Europe of the Middle
Ages where leprosy and the Black Death were declared to
be God's punishment? Death from earthquakes and volcano
eruptions too. And today? When hurricane Katrina flooded
New Orleans, weren't there otherwise respectable people
who declared this natural disaster to be God's punishment
for the immorality of the Big Easy?

Mino knows all this, but Huo has obviously started to
take advantage of his despair. He may be trying to convince
Mino that his karma has already been cleansed thanks to his
energy treatment, causing the GBM to shrink.

Who is Frank Huo? Does he have a hidden agenda, or is
he himself a victim of a belief system that has been promoted
for such long time that it has already become part of the
subconscious, almost part of common sense?

Frank is still standing behind me. "There are many Chinese restaurants around here," he says, ripping me out of my
thoughts, "if you haven't had dinner yet."

Dinner? I had no breakfast, no lunch. I thank Frank and
ask him to leave.

I lie down beside Mino in his bed and watch him breathe.

The left corner of his mouth is drooping. His skin is pale,
almost gray. I can see how grave his condition is.

Friedemann arrives. I whisper to Mino: "Papa has come."
Almost imperceptibly he nods, and there is a shadow of a
smile.

We decide at once to bring Mino to the Columbia University Hospital. He can barely walk to the elevator and out
to the taxi. We have to hold him from both sides.

In the taxi, Mino rests his head on Friedemann's shoulder.
His iPhone rings. Mino takes the call. It's Dr. Wecht's nurse
in Pittsburgh. In a flash Mino's voice sounds decisive again
as if nothing had happened. As soon as the call is over, Mino
slumps back, resting his head on Friedemann's shoulder. A
striking contrast. Looking at Mino's miserable condition, it
is hard to comprehend how he was able to speak with a voice
so full of energy just a few seconds ago.

At Columbia, after filling out papers and more papers,
going from one station to the next, answering the same
questions over and over again, getting blood pressure and
temperature measured countless times, Mino is finally
allowed to move from his wheelchair to a bed—a kind of
promotion in view of the severity of his condition.

After the bureaucratic ordeal, several doctors come, one
after the other to ask questions about Mino's medical history.
Mino engages them with counter questions, which lead to
spirited discussions about GBM, its possible causes and possible remedies. Mino shows no trace of fatigue. He is alert,
sharp, and even cheerful, as in his normal days.

However, as soon as one doctor leaves and before the
next one comes, Mino sinks back into his pillow, half asleep,
weary, and pale.

I am amazed at Mino's rapid transformation from obvious fatigue to full concentration and focus. He changes from one state to the other within a blink of an eye and reverts back to being as lethargic as before. But his will is unbroken, as is his never-ending thirst for understanding the things around him. I look at Friedemann from the side and see in his eyes the same sense of wonder.

Shortly before midnight Mino is brought to the MRI unit, a place with freezing cold air conditioning. Should take half an hour, we are told, but we have to wait until two A.M. for Mino to finally come out. A nurse tells us that according to Dr. Bruce's instructions Mino has to go immediately into the Acute Care Unit for constant observation.

While the nurse wheels Mino's bed down the long corridor, we run after her. She turns around. "We will take good care of him," she assures us, and waves us off.

Back in the hotel at three A.M., Friedemann and I instantly fall into deep sleep. At seven A.M. Mino calls. "Nurses come in and out," he says plainly, "all night.... Put lights on and off. Hook me up for EKGs.... A dozen electrodes or so. Draw blood every two hours. Impossible to sleep."

"Have you heard from Dr. Bruce?" I ask.

"Yes, he came at six. Just left. My tumor has grown to seven centimeters in diameter. More than triple in volume. According to him I have no option left. Surgery." After a long pause, Mino adds: "I need a quiet place.... I need to find myself.... Can't do it in this lousy place. I want to get out of here."

Friedemann and I hurry back to Columbia. As we walk down the long corridor to the Acute Care Unit, Dr. Bruce happens to come out of one of the patients' rooms, tall and slender. "I saw Mino this morning," he says in his unhurried way "Talked to him for forty minutes. His condition is very grave, but his mind is stunningly clear and sharp."

We walk through the open door into Mino's room. I see

him lying on his bed, eyes closed, hooked up by many tubes to some machinery left and right with blinking lights and occasional beeps. He seems to have dozed off. Friedemann and I exchange glances. We tiptoe to two chairs along the wall and sit down.

I stay with Mino all day and all night, witnessing the constant coming and going—nurses and nurse assistants, neurologists, neuro-oncologists, and residents. Friedemann leaves to look for a nearby hotel where we could stay for the time Mino will need to recover from the surgery.

Towards the evening of the day before the scheduled surgery, Mino still says: "I want to get out of here and go somewhere quiet."

"You can't go out anymore, Minochan."

"I need silence."

"You can bring silence here, Minochan, through power of meditation. Think of the redwood trees you love so much and of all the places you remember, the Grand Canyon, the Swiss Alps, the hot springs in Japan, the energy of the universe— you can bring all here into this room. Minochan, please."

I go out into the hallway and stand in front of the door to Mino's room to make sure that nobody comes to disturb him.

Almost an hour later Friedemann arrives, and we both go in. Mino wants to stand up and look at the sun. We support him from both sides and step to the window. A magnificent evening sky spreading over the Hudson River with the George Washington Bridge to the right. Mino watches the sun setting.

After a while Mino turns around, a strange calm on his face, a resolve, even a smile. No agony and no uncertainty. We bring him back to his bed. He quietly signs all hospital papers, what to do in case of complications. Yes to a feeding tube. Yes to a breathing tube. Yes to resuscitation. No to organ donation.

"I want to shave—to be clean for the surgery," he says.

Back in his bed, Mino starts checking emails and writing messages.

"I've set up this workshop," he says as if we knew it.

"Which workshop?"

Mino doesn't answer. He continues typing vigorously with his right hand, his left hand hanging limp down the side of his bed.

After a while: "On my nanosat mission...two hundred nanosats, at any time two in line of sight, mostly three, from any spot on the surface of the Earth." He glances at Friedemann as if to see whether he is following. "Two hundred nanosats is the minimum to get a fully connected system."

Friedemann nods hesitantly. "How do you connect them?"

"Lasers."

"But this has never been done before...except between just two satellites."

"So what?" Mino quips. "If it hasn't been done before, all the more reason to do it now." Defiantly he repeats what he has said before, that the folks at NASA Headquarters still don't believe it would be technically possible to operate a constellation of two hundred or more satellites, with laser-based cross-communication. A massive amount of information would be transmitted between the satellites and to the ground.

"With lasers...we can transmit megabytes per seconds, even gigabytes," Mino continues to type with his right hand, sending messages to everybody whom he wants to attend the workshop. "See you in Manhattan...after my surgery...exact date and place to be determined."

The night at the Columbia hospital can be terrifying. Different staff for the night shift. Many are hired through staffing agencies, literally off the street, with little regard for their competency.

During the night, just after midnight, Mino develops a shortness of breath. It becomes worse by the minute. Probably a reaction to Percocet, a pain killer that the hospital gave him that evening. I see the oxygen saturation value on the monitor dropping from 98% to 95...90...85. Mino gasps for air. His skin is turning ash gray. I pull the cord to the emergency bell and hear it ringing. No nurse. I run out into the corridor. At the far end a nurse appears and starts leisurely walking towards me, a short, fat woman. By the time she arrives, the monitor values have dropped to the 70s.

"He needs oxygen...oxygen."

"Whaaat?"

"He needs oxygen ...quickly...now."

"We ain't have no oxishen," the woman says.

I point to the oxygen mask dangling from the wall.

"Can't touch. Must call a dactar first."

Mino's oxygen monitor is down to below 60.

I storm out into the corridor, screaming for help at the top of my lungs. Another nurse comes, running. She takes a quick look at Mino's monitor, which stands at 55. She rips the bright yellow oxygen mask off its holder, turns a valve, and slams the mask onto Mino's face. For several minutes she remains in her bent-forward position, pressing the mask against his face and watching intensely as Mino starts to breathe normally again.

"That was close," she says, as if talking to herself, while adjusting some nobs on the monitors that never stop blinking, "Fifty-five.... Never seen a value that low."

Next morning, on the day of surgery, Dr. Bruce comes again at six A.M., unhurried, with a heart-warming smile. After him, a lovely middle-aged Indian nurse, who looks at

Mino and me with affectionate eyes. I soak up her warmth after a sleepless night.

Mino quizzes Dr. Mark Otten, Dr. Bruce's resident, about the exact surgical procedure.

"Everything is set up as you have requested," Dr. Otten explains. "You'll see the LCD panels along the walls with all your MRI scans, the stereotactic instrumentation, and the surgical saw for cutting through the skull."

"How large a cut?" Mino asks. "How large a hole?"

"No hole, just a straight cut, about two inches long. After this procedure, we'll wake you up for testing the cortex. After that, we'll sedate you again for the rest of the procedure."

"No general anesthesia?"

"No."

Dr. Otten speaks as if he were teaching a science class. Mino listens attentively, as if he were not the object of surgery.

After Dr. Otten leaves around seven A.M., Mino looks content. He leans back into his pillow and slides into sleep, holding my hand, at peace with himself, breathing deeply, sometimes even snoring a little bit. Friedemann and I, sitting on either side of his bed, watch him for almost two and a half hours.

Around twelve-thirty two nurses come and roll Mino's bed to the surgery floor. We follow. In the preparation room the chief anesthesiologist explains what he is going to do. The surgery would take four to five hours.

As his bed is being pushed into the surgery suite, Mino looks at us with a mischievous smile and says in Japanese, "Don't worry...see you soon."

We try to relax. Friedemann falls asleep. I can't. I stare at the clock on the wall. Exactly two hours and ten minutes later, a nurse in green OR outfit comes hurriedly. "Dr. Bruce wants to talk to you," she says out of breath. "Please follow me."

I jump up. Something horrendous must have happened....
Dr. Bruce wants to talk to us...in the middle of surgery....
Something horrendous...

Friedemann, still drowsy, runs after me.

In front of the door to the surgery suite we have to wait.
A few minutes later, which seem unbearably long, Dr. Bruce
comes out, still wearing his surgeon cap and in green OR
coat. I try to read the expression on his face.

"Well," he says with a smile, "everything went smoothly,
even exceeded my expectation. It was a good surgery, even
a very good one. I was able to take out more than 99% of
the tumor."

SILVER LINING

I'm back. My tumor is gone. My wound has healed. When
I pass my fingers through my hair, I can feel the scar, but
barely. I have regained full use of my left hand and left leg. I
can walk again. I can type with both hands on my computer
keyboard. I can drive. I'm full of energy.

Deb Feng comes all the way from California to see me.
Such a surprise. "You haven't changed a bit," she says,
unpacking a rich dark chocolate cake she has brought with
her. "Very little sugar," she points out.

My technical meeting in Manhattan went well—very well
in fact. Everybody who attended confirmed that it was an
important meeting. Ten participants, from the New York
area and Washington D.C. We went through my two hundred
nano-satellite mission concept and discussed many of the still
open questions of what instruments to put on each satellite,
how to design the orbits, how to launch them and how to
keep all two hundred nanosats together in a tightly controlled
constellation over long period of time.

Using laser communication.

Much of the discussion was, of course, dedicated to this

critical component of the laser communication. This has never been done before—not at that level of complexity and precision. How can the nanosats communicate with each other and how can they talk to the ground? How can we keep them in formation with millimeter, even micrometer precision? Maybe we'll need two or three larger satellites in higher orbit as relay stations for handling all that enormous data flow, receiving and downlinking to ground. We'll have a very large amount of data. We need to consider onboard computing to reduce the data size. Quite a challenge…but my calculations show that it can be done. Brian Hibbeln had some good ideas how to fund the project and promised to look into it.

Now, after flying across the North American continent and over the Pacific, Saya and I have arrived on the Big Island of Hawaii. Her friends in Tokyo invited us to stay in their mansion in Kona as long as we want. The house is high up on a slope with a commanding view of the ocean and the sky. What balmy air! All day quiet except for chirping birds and the faraway sound of breaking waves. The sun is bright but not burning. I can be in the garden all day for meditation.

I sit on my large purple-colored cushion and close my eyes, breathing deeply and ever more slowly. I feel all the tensions and anxieties of the past weeks in New York peeling off little by little. Here is where I wanted to be and this is what I wanted to do.

The decision to fly to Hawaii was not easy. The reason is that, within barely six weeks of Dr. Bruce's operation, my GBM came back, forming a new tumor in the same spot in my brain. All this despite the fact that Dr. Bruce had taken out 99% of the cancer cells and that the MRI immediately after surgery had shown on my right parietal lobe nothing but a large fluid-filled cavity.

Dr. Bruce offered me a second surgery, which he said

would be much simpler and easier than the first but still with the frightening certitude that he'll again not be able to remove all of the cancer. Even if I submitted to the second operation now, the GBM would be expected to come back. I must fight it by different means.

Over the Thanksgiving weekend I decided, after intensive discussions with my parents, against a second surgery. The foremost reason was that I wanted to continue my alternative path of fighting the GBM by other means. In fact, the immune response in pathological samples from Dr. Bruce's surgery had shown that the immune defense of my body was very strong, as evidenced by CD4 and CD8 markers inside my tumor. An even more striking result was that there were P.53 markers in the endothelial cells lining the blood vessels, indicating that my body is able to recognize the GBM cells and mount a strong immune response. If I had not already devoted my efforts to alternative methods, I would not be in such good shape. I have to continue on this path.

This is why I had refused to submit to radiation and chemotherapy after Dr. Bruce's operation. "Gold Standard of Western medicine," is an encouraging name, but radiation and chemotherapy wreak havoc on the immune system. My immune system is working quite well, surprisingly. At the same time, I know that the GBM has a scary ability to order the body to supply it with extra blood vessels. They bring nutrients to the tumor cells, allowing them to thrive. If it were not for this dense network of very fine blood vessels, the tumor would not be able to amplify. Growth of the blood supply is known as angiogenesis. Normally angiogenesis is wonderful to have. When a heart doesn't work properly due to a lack of capillary blood vessels, angiogenesis is a savior. In case of a brain tumor, however, angiogenesis is the worst enemy, because it sustains the tumor by feeding it with glucose, which the tumor cells convert by fermentation, not by oxygenation, enabling its perpetual growth.

To stop my GBM, I have to find something that fights the angiogenesis. Plus, I have to find it fast—really, really fast.

One of the formidable drugs in the Western arsenal is Avastin, which has the ability to limit the blood supply to tumors. Avastin, however, is also considered a "drug of last resort." Most often, after a while, the cancer cells mutate to become resistant to Avastin. When this happens, new tumors spring up in many places in the brain and proliferate. Avastin is unable to stop them.

Another ghastly property of GBM is that the cancerous glio cells infiltrate the surrounding healthy brain tissue. Removing them all by surgical means is not possible, because, undetectable by MRI and by any other techniques, they already have penetrated the surrounding healthy tissue. This is why no surgeon can take out all of them.

I have read so much about it, because the argument can be made that my GBM tumor could have been totally removed, if only I had submitted to surgery at the earliest date, when it was discovered back in Palo Alto in August 2009. Unfortunately, this is not true.

The deep penetration into the surrounding brain tissue is a predicament that exists with any GBM, regardless of size. There is never a time when a GBM can be removed with all its far-reaching tentacles. It's just impossible for any surgeon to cut into the brain to remove those billions of GBM cells that are still invisible and untraceable.

There is, however, a silver lining. Our immune system is not entirely unable to recognize cancer cells. The problem is that the soldiers of the immune system are not quite sure when and how to launch an attack. Those cancer cells are, after all, cells from our own bodies; they look like our healthy body cells, except that they have gone out of control. They have reverted to a stage similar to those of the earliest forms of life that populated the Earth long before there was oxygen in the Earth's atmosphere.

At that ancient time, some three billion years ago, there were only single-cell organisms; and they all had to survive and thrive on fermentation processes, just as the GBM cells are doing now in my brain. Having converted back to this archaic, single-cell existence, the GBM cells have lost the capacity to follow the instructions of our body to organize as functioning cells. Instead, they spread and multiply in an unorganized manner, using the surrounding healthy cells as the substrate from which they can extract nutrients. In addition, they have that insidious capability to persuade our bodies to supply them with plenty of blood vessels, so that they can be even better at extracting the nutrients for growing and growing and growing.

If only I could command my immune system to stop that nonsense.

If only I could help those soldiers in my arsenal of immune cells to recognize those nefarious cancer cells.

If only I could train them to become better fighters on my behalf.

If only I could educate them to go after every single GBM cell, billions of them, grab them, devour them, and clean them out of my body.

After deciding against another surgery, I have to do two things: first, fortify and invigorate my immune system by alternative treatments, and second, find something safer than Avastin to fight the angiogenesis.

This is why, despite a small recurrent tumor inside my brain, we all agreed to leave a windy, icy New York. We could have gone home to Northern California, but there it is cold and rainy in December and January, impossible for me to sit and meditate outside in the sun.

Usually meditation is something to be done inside, in the stillness of a room or in the serenity of a church, a temple, or maybe in the garden in the shadow of a tree. But meditating for hours facing the sun? My meditation requires just that: to let the power of the universe flow into me, so that I gain tenacity and calm.

So, I sit here on my large, purple-colored cushion, meditating under the Hawaiian sun with plenty of water to drink, and after sunset I stay in the garden, listening to the sounds of the twilight hour.

Many fond memories of the Big Island go through my mind. They include my deepest scuba dive, which I did along the Kona coast down to 267 feet, nearly 90 meters. Two enormous manta rays swam around us for a few rounds, curious to see strange, bubbly folks at such depth. My diving instructor and I estimated that these rays had a wingspan of about seventeen and twenty-five feet. Staying ten minutes at ninety-meter depth had to be followed by two hours of helium-enriched decompression—think of me as a diving Donald Duck.

I also remember the astronomy conference in Kona in January 1998, when the El Niño caused especially dry air and created some of the most spectacular viewing conditions on top of Mauna Kea. One night some of us drove up to the midway station. We rested there for an hour to acclimatize and then went on to the top for observations on the Caltech Submillimeter Observatory.

I still remember seeing, with the naked eye, stars that can hardly ever be seen from the ground. I saw Polaris, a magnitude 9.5 star, difficult to spot among its many companions. I saw the gorgeous Andromeda nebula, which is almost 200 arc minutes in width. I saw all the way down to the Southern Cross, where there are planetary nebulae in intricate red-green-blue-white glory near the plane of the Milky Way.

I think of my early visit to the Radio Telescope in Effelsberg, near Cologne, with our Computer Club. There I saw for the first time, shrouded in the interstellar dust clouds of the Milky Way, the birth of stars. Now, on top of Mauna Kea, I looked through its Infrared Telescope, zeroing in on the latest supernova explosion in our galaxy—the death of a star.

ANGIOGENESIS

I have been doing extensive research to set up my treatment plan, choosing from about four hundred cancer fighting protocols. Included in the list are typical pharmaceutical drugs like decadron, keppra, an anti-seizure medication, and nexium to protect my stomach from the effects of decadron. In addition, I have a long list of natural drugs to fight angiogenesis. One of them is from China, is approved for brain cancers, and claims to penetrate the blood brain barrier. Also, sutherlandia, a South African herb, together with oleander extract. All in all about twenty different medications and supplements, to be taken in about sixty doses daily.

Saya has arranged all my drugs and supplements on one side of our large dining table. She is worried that I might miss one or another drug, as I'm often absorbed in meditation or in my research on the medical front.

Over lunch and dinner, we chat and chat and laugh as we so often do, looking at the sky and over the ocean, noting how the light changes, as do the colors and the sounds.

We talk about my physician here, Dr. Tony DeSalvo, the only oncologist serving a community of 50,000. He and his staff of about ten treat 350 patients every week in two clinics, one in Kona, the other in Waimea, and he personally sees every one of his 350 patients. He is a rather heavy-set guy in his mid-forties and by far the most caring doctor I have ever seen. Though he is not a neuro-oncologist, he is broadly knowledgeable and even-tempered. Dr. DeSalvo

writes up decadron and another MRI for me. To get prescription drugs we drive to the nearest Long's pharmacy. The
people there are utterly disorganized or relaxed in a Hawaiian
way, whichever best describes their dysfunction. They never
have the item they promise the doctor's office. They are waiting for a delivery from the mainland or from their center in
Honolulu. A few days' delay seems to be part of life here.
Aloha.

Compared to the 1990s, the traffic on the Big Island
has become painfully jammed with more red lights, despite
stretches of four-lane highways. Small stores have all but
disappeared and in their stead gigantic chain stores are popping up all over.

To get a high-quality MRI, we have to drive for an
hour and half on approximately forty-five miles of narrow,
winding two-lane roads to the Waimea North Hawaii Community Hospital. My latest MRI indicates that my tumor is
still growing, but the growth rate has slowed. Quite significantly. The volume doubling time has gone from nine days
to twenty days. That may not sound like much, but it is an
exponential growth, meaning that instead of quintupling in
volume every three weeks, it now only doubles every three
weeks. The surface area growth is constant, which means
that the angiogenesis has at least leveled off. This gives me
the confidence I need to move forward on my path.

Looking at the biology and chemistry of my cancer, I've
concentrated until now on getting the growth rate down by
controlling the angiogenesis. This strategy seems to be working, finally. Cancer, however, is unpredictable, and the growth
of the tumor is not constant, but can become chaotic and
riotous. I have to deal with it. Effectively I'm my own doctor, which means a lot of research and a lot of learning. The
weight of the work is on my shoulders.

I have to watch out not to fall for information based on

statistics, because most of it, as used by the medical community, involves frequency statistics on multivariable data sets. Its math was developed before the French Revolution. Pseudo-math, I would call it. The medical community still sticks to the old statistical dogma, though it has drawn harsh criticism from many physicists including the Nobel Laureate Phil Anderson in the mid 1980s.

Any drug has to have a scientific basis, with reliable case studies that I can verify. For that I have to find the relevant papers, download them or request them from doctors, read them, understand them, and evaluate them. Only if they stand up to all that scrutiny will I consider adopting them as part of my regimen.

Once I was alerted to a newly FDA-approved anti-angio-genesis drug that has to be given intravenously for three hours every day over a period of sixty days. I discussed it with Dr. DeSalvo. He agreed to do the procedure in his praxis. He gave me a prescription and I almost ordered it for delivery by Tuesday, December 22. However, after a day of all-consuming research, I decided against this new drug, because there were just too many unknowns and potentially fatal consequences. More research remains to be done. I can't relent, until my tumor starts to shrink or at least stops growing.

I have received many, many emails from colleagues and friends at NASA as well as at other federal agencies offering me their leave time under the Voluntary Leave Transfer Program. To all I would like to express my deep gratitude. Many have helped me with no expectations of anything in return. They are supporting me through their actions, with their thoughts, and with prayers. I'm filled with thankfulness.

Some seem to be offended for not getting any response from me, however. Since August 25, 2009, every single day I have only been a short distance away from major surgery or death. Please remember that for a cancer patient with a fast-growing brain tumor, nothing else counts but fighting for life. I have to be egocentric, because my life depends on it.

FIREWORKS

Tonight the fireworks are going off at the resort hotels down the slope along the coast.

Welcome to 2010.

I'm enjoying a great dinner with Saya, hearing continuous explosions far away. Papa is at home in the Bay Area, recovering from the AGU meeting in San Francisco and working with his unrelenting and infectious enthusiasm.

Roaring fireworks remind me of a story which Saya completely forgot.

Well, years ago, back around 1993 or 1994, when we three still lived in San Francisco and I was commuting over the Bay Bridge to the Physics-Astrophysics Department of Berkeley, we decided to go and watch the Fourth of July fireworks. There are ferry boats leaving from Fisherman's Wharf out towards Alcatraz Island. From there the view over the fireworks that San Francisco puts on every year is fantastic. But beware, floating on a ferry in the Bay can be very windy and frosty.

For some reason, we were late and came up with a last-minute plan to find parking somewhere close to the Fisherman's Wharf and buy tickets for a ferry ride. We wanted to find a nice restaurant and have dinner before hopping onto the ferry.

However, there was just one little problem: many people had more or less the same idea. As widely known to San Franciscans, parking near the waterfront on July Fourth is not a trivial undertaking. Luckily, we found a spot somewhere on Bay Street, and I was happy that step one of our plan was executed flawlessly. We also bought the tickets for the ferry. Excellent so far.

Now for dinner. Wherever we looked, the waiting time to get a table in a restaurant was measured in hours—everywhere.

So, we improvised. We set our eyes on a grocery store on Beach Street, pretty empty, offering us a large selection of foods to take out. We decided on freshly roasted chicken, a three-feet long French baguette, some ham, Swiss cheese, and salad, not to forget hot tea in paper cups and plenty of napkins.

Now the next question was where to eat. Everywhere was full of people, including the coffee shops. Then we saw a wooden bench, with the last of the sunlight warming it, wide enough to sit there with the food and the drinks beside us. We put our grocery bags down by our feet. Knowing how cold it gets in San Francisco, we had dressed warmly, meaning we wore several layers of clothing. Being from the thrifty post-WWII generation, Saya hates to throw anything away, unless it really falls apart. We were all very hungry.

So, imagine for a moment three PhDs, in worn-out clothes, totally color mismatched, and a patched blanket, with disheveled hair, sitting on a wooden bench along the sidewalk, eagerly munching away, with grocery bags in front of them. My parents were engaged in some spirited discussion, as they almost always are, while strangers walk past.

Nothing remarkable so far, but then things got interesting. A man walked up to us, mumbling something like "Happy Fourth," and he dropped a one-dollar bill into one of the now empty grocery bags. Someone else threw in a few quarters about thirty seconds later. Soon all kinds of strangers dropped more one-dollar and five-dollar bills in our grocery bags. And many coins. By the time we left the wooden bench to head to the ferry, there was enough in our grocery bags to cover our dinner and our ferry tickets.

Happy New Year from Hawaii!!

PLACE OF REFUGE

I was at Honaunau Bay the other day for about five hours
of meditation in the sun, overlooking the coast. No sandy
beaches here. The coast is rough with black lava flows jutting
out into the ocean.

The place used to be a quiet paradise with a Royal Palace and a "Place of Refuge." Built on the thousand-year-old
lava flows, both played an important role in the history of
Hawaii. The Place of Refuge was considered sacred with, its
own priests, called kahunas. It was a sanctuary for those who
had broken any of the many taboos of Hawaiian society or
had been defeated soldiers from the many feuds. If they could
manage to reach the Place of Refuge, they were safe.

Hearing this story reminded me of the asylum in ancient
Greece, where certain temples, sacred groves, and statues of
the gods were considered to possess the power of protect-
ing those who fled to them for refuge, including runaway
slaves. Asylum seekers could find refuge in the temple of
Poseidon in Akron Sounion, at the Altar of the Twelve Gods,
and in the Temple of Artemis in Ephesus. These sites were
sanctuaries like the Place of Refuge at Honaunau on the Big
Island.

The sad thing is that, by now, the beautiful Honaunau
bay has become too famous. As a consequence, the place is
overrun with merrymakers and tourists, intermingling with
divers and snorkelers. Motor boats race past Honaunau at
high speed, splashing and roaring.

Back to the rough black lava rocks. Those in the tidal
zone have small pools, often filled with crabs, juvenile fish,
and various sea urchins. Some tidal pools are connected to the
open ocean by underground tunnels. On windy days, when
the waves break way out at the edge of the lava flows, water
in these pools often shoots up to form spectacular ten- to
fifteen-feet-high geysers.

I remember something that happened at a similar place thousands of miles away along the Pacific coast of Japan. There the rocks are also made of black lava. Those in the tidal zone also harbor pools. In one of them Grandpa and I found a large sea bass, longer than my outstretched arm and glowing in a magnificent dark-reddish color. In most cases, when a fish is caught in a tidal pool, there is no real reason for alarm, because it can wait until the tide comes back. But there, at the coast of Japan, at the peak of summer, the situation was different. That tide pool was very small, more like a kitchen sink, and its water was getting warmed by the afternoon sun. Furthermore, the tide was still going out, and we could see that the water level in the pool was sinking. This sea bass was going to have a painful death. In addition, there were men walking up and down the coast, looking into tide pools. Many were carrying buckets full of fish.

As I approached the fish in its small tidal pool in Japan many years ago, it got very agitated, trying desperately to hide. I kept motionless and waited for it to calm down, talking quietly to the fish. Then I approached the pool, and slowly lifted the fish out of the water. I continued talking, while I carried it to the ocean's edge, careful not to slip and fall into the water myself. I gently set it free in the water. The fish stayed motionless at first, then circled in front of Grandpa and me, jumping out of the water almost half its body length, darting here and there. Then, after a final jump, it disappeared in the depth of the ocean. The anthropomorphic interpretation, of course, is that the fish was full of joy. For reasons I will explain next, I'll stick to that version.

In the intervening years I have noticed that when I walk along a stream or a river or the ocean's edge, fish often congregate close to where I am. Hard to believe. Sometimes, when this happened, not far from me there were fishermen with rods and hooks. They had caught no fish as far as I could tell, even though a dozen or more fish were drawn to me, as if they knew that I posed no danger.

When I lived in Dayton, Ohio, I used to go to the Glen Helen Nature Preserve, a special place in my spiritual journey. A stream with very clear water runs through the Preserve. There are large river carps, known to be quite shy, but when I stood at the water's edge, they came. They would not move away when I crouched down. They even let me pet them. In a creek that joins the stream, minnows would congregate in large numbers wherever I walked, and the water would sometimes seem to boil with them. When Saya visited me in Ohio and came to see the river carps and minnows in the preserve, she could see them from a bridge, but when she came closer they would swiftly move away, hiding under some green waterweeds.

I was reminded of all this at Honaunau. Fish are here and so are fishermen. Nothing in their buckets even though they are trying to catch for hours. At the same time, standing on the lava flows, I could spot several large fish congregating towards me. Thus, I'm inclined to believe that the bond between fish and me is unbroken.

Modern scientific research has shown that humans are not as unique in many ways as we are often taught, especially when it comes to feeling pain, joy, or grief. Much brain research on the endocrine system has shown how much animals can experience pain, joy, and grief. The monkey mother, who carried her dead baby for days in her arms. The elephants that gathered around a dying or dead family member. The killer whale mother somewhere along the New England coast whose baby had died a few weeks after birth. She carried its body for nine days while traveling with the pod up and down the coast. Don't tell me that animals don't feel pain and grief.

Many animals also have what we may call preferences for one or other individual, all the way down the evolutionary chain—from mammals, to birds, to other vertebrae like lizards, snakes, amphibians, and fish. Even some highly evolved invertebrates, such as octopuses and squids, seem to have individual preferences. They not only recognize other individuals of their kind, but they also like to "hang out" with their "best friends." I suspect that sometime in the not-so-distant future someone will show that love and longing as expressed in Shakespeare's plays are as relevant to a considerable part of the animal kingdom as they are to us. Taoists have always recognized the universality of pain, joy, and grief—even longing and love—beyond the human realm.

In light of all the science we have on life, the question burns in my mind: why do humans hunt, catch, and kill animals? The biological answer is that nature has made us an omnivorous species that thrives best on a diverse diet derived from both plants and animals. Therefore, there is a necessity for humans to kill animals for food. What I have in mind, however, is the deliberate killing of animals for pride and pleasure. Why do people get excited about "bagging" a large deer or antelope, a lion or a tiger, or a large fish for that matter? Or herding the dolphins into a bay like the one in Taiji, Japan, from which there is no escape. There the local fishermen take great pride in killing dolphins by the thousands, more than twenty thousand every year—a slaughter orgy that lasts for days and nights on end. The locals and even the Japanese Prime Minister justify this bloody business as an age-old Japanese tradition. But it is by no means a tradition and not age-old. The earliest reports of the dolphin mass slaughter date back to late 1960s, not before.

Why do people do it and draw enjoyment out of it? First, people do it because they can or because they look at it as a trophy sport, which makes them feel great. Second, to put it in crasser terms, the movie "Natural Born Killer" comes to mind, where killing is entertainment for the debased instincts of humanity. Killing as entertainment has not stopped since the gladiator days in the Colosseum in classical Rome.

Sitting at the edge of black lava fields at Honaunau on a quiet afternoon, surrounded only by the sounds of waves and wind, I close my eyes. Time is happening on so many different scales: the life of fruit flies that live only a day, the small trillium flowers that last a few weeks, a myriad of small insects that live in the forests' floor for several months, the cicadas that spend seven or more years in the soil and emerge for only ten days to mate and reproduce, the sea turtles that may live up to a hundred years or more, the oak trees and other trees that live for centuries or the redwood trees that live for one or two thousand years, even longer, reaching enormous girth and height. In fact, redwood trees never die, because when an old giant redwood falls, new young ones sprout from the same root system...for ten thousand years and longer. That's why redwood trees are called *sempervirens*, "forever living," It is as if "life-time" on earth has become multicolored or elastic—happening on scales from seconds to millennia. And considering the cosmos, we find that Earth's limits of life-time are far surpassed by the stars, whose life cycles are counted in billions of years.

Looking over the smooth black lava fields and the ocean beyond, I feel so small. I think back to the time, hundreds, even thousands of years ago, when this was a place where red-glowing lava from the mouth of the Kilauea met the ocean, producing steam as it plunged into the water.

I think of the time when the seafaring Polynesians discovered the Hawaiian island chain and chose the Big Island as the site for their kings and royals. I think of the sanctuary they built here for all who were fleeing persecution.

I came here, I guess, also to flee from persecution, the

persecution inflicted on me by the GBM that is still growing in my right parietal lobe. I wish I could stay here all night, sitting on the smooth lava flow, still radiating the heat of the afternoon sun.

As the sun settles below the horizon, the sky turns from blue to red to gray, and then so deep azure that the first stars begin to sparkle. I am carried back in time. I hear voices from inside the sanctuary welcoming a refugee who has reached safety. Waves and winds join them.

Ever more stars show on the sky, including many of my stars that I know from the time of our IRTS mission, which gathered the light from tens of thousands of stars shining in the infrared part of the spectrum, invisible to the human eye.

I cannot ask those stars to help me in my fight against the GBM. They are too far away and the light I see now, the light that our IRTS mission recorded, had left those distant points of light thousands to millions of years ago.

All I can hope for is the sun to help me strengthen my *chi*, that profound force within ourselves, which is linked in some strange but believable way to our immune system and its capability to fight whatever sets out to harm us.

It's a very tough fight, because cancer cells know how to evade detection by the soldiers of our immune system. At the same time, the cancer cells cannot totally hide who they are and where they are. If I'm able to control the angiogenesis, to cut off the tumor from the ample nutrients provided by capillary blood vessels, I'll be able to win my fight.

Saya comes and says, "We must go. The gate will soon close."

Turning away from the ocean, we gingerly find our steps back in near-total darkness, over the lava flows to the sandy trail between palm trees that leads us to where I parked the car.

WHEN I WIN MY FIGHT

February first, 2010 was a turning point for me. From that day onward, my condition began to deteriorate. Before that day, I was fully functional, driving our rental car, moving effortlessly everywhere, even running over the lava fields in Honaunau, cooking my favorite food at home with Saya, doing research on the computer, and sitting in meditation. After February first, I was going downhill, fast.

An emergency service brought me to the emergency room of the Kona Community Hospital, where I received intravenous infusion of decadron to reduce the edema caused by the tumor. Soon my headache subsided, but I knew it wasn't possible to rely on decadron for a longer time. Decadron is not a solution.

Saya must have called Papa in San Francisco, because he flew into the Kona Airport within an amazingly short time of just six or seven hours. Both of them were with me in the hospital. Dr. DeSalvo, my oncologist, told us that treating my condition was beyond his capabilities. He urged my parents to take me as soon as possible to the University of California San Francisco Medical Center and its UCSF Neurosurgery Clinic.

I refused to follow Dr. DeSalvo's advice. I refused to follow my parents' plea, not because I was afraid of a new surgery, but because my own strategy against angiogenesis seemed to be working and my meditation in the sun and in the ocean breeze seemed to strengthen my immune soldiers. I was clinging to the hope that they would attack and devour the GBM cancer cells. I wanted to continue my path in Hawaii.

Just a few days before February first, I woke up early in the twilight and saw the full moon setting in the Pacific Ocean. The contrast between the pale twilight and the magnificent full moon sinking into the ocean was so beautiful that tears streamed out. I was happy, really happy for being alive.

I wondered who among my favorite painters would have been best at rendering that fleeting moment on the canvas: Claude Monet, Childe Hassam, J.M.W. Turner, Vincent van Gogh, or Rembrandt.

When I'm healthy again, I want to return to my drawings and develop my watercolor skills. I'll visit exhibits again as I did so often in Cologne, in Zürich, and in Washington, D.C. during my time at the NASA Goddard Space Flight Center.

When I get rid of this stupid tumor, I'll continue my project to fly a constellation of nanosatellites, two hundred of them, all transmitting among each other and to the ground by laser communication technologies. This has never been done before, at least not on that scale. I have been pushing hard for this ambitious mission, at NASA Ames, at NASA Headquarters, and so many other places.

"It can't be done" is what I hear most often. "Technically impossible."

"Too far ahead of time" is the friendliest comment.

It's ahead of time. That's why I'm doing it. Doing something that others can do as well is not what attracts me. I like the seemingly impossible. How else can we have progress?

A big satellite costs hundreds of millions. The nanosats that I have in my drawing board can be built and launched for much less, probably as little as a hundred grand.

More important, if we build and launch them for a fraction of the price of a billion-dollar mission, and if they can do about three quarters of what a billion-dollar satellite can do, just think of all the good things that would come out of my two hundred nanosat constellation. I don't even want to start counting the benefits for the world. I can put cameras on them that look for wildfires. With two hundred of them up there, we'll be able to revisit every point on the surface of the Earth every fifteen or twenty minutes. The key to fighting wildfires is early detection, so that the tanker airplanes get quickly on their way to extinguish the fires with their load of fire retardant while the flames are still small. This is just one example, where a large number of nanosats in the sky can fulfill a very important function.

Yes, there are downsides. My nanosats might be used for spying. Everything in life, everything including technology, has downsides. Alfred Nobel knew this when he saw that dynamite, his invention, was being used for evil as well as for good. That's why he set his fortune aside to create his Nobel prizes—all for peaceful applications, because evil applications will be done anyhow.

I trust that my nanosats will be the future. I trust that they'll help Papa cracking the code of all these strange and seemingly incomprehensible signals that the Earth produces before major earthquakes.

Papa has such a hard time with the old guard among the seismologists, who proclaim categorically that earthquakes cannot be predicted. They say so because they, the seismologists, haven't been able to get even close to this elusive goal. But Papa is well on his way. He has solved the riddle. He has worked long and hard on deciphering the underlying physics. I checked his physics. It's okay, actually better than okay. It's new physics. That's why the old guard is against it. As soon as I get rid of my nasty tumor, I'll help Papa.

It's not so different from what I have been going through with my idea of a two-hundred-nanosat constellation. I had to travel so often to NASA Headquarters, to the NASA Goddard Space Flight Center, to Marshall, to Stennis, to Houston, to JPL, even to London and Tokyo.

The good thing is that Pete Worden, the NASA Ames Center Director, is fully supporting me. Al Weston, our Project Director, has come along on some of my most tiresome travels. I think I have turned the corner on my nanosat project. That's why this workshop in Manhattan was so important, which I had convened just ten days after my surgery at Columbia.

I must get back into the trenches.

When I get rid of this nasty tumor, I'll also push ahead with my neuroscience project, which I had started before the GBM sidetracked me. To use nanoparticles to deliver drugs to brain tumors. A new field, still wide open and full of opportunities. Steve Zornetzer at NASA Ames, himself a neuroscientist by training, strongly supports my ideas. I have to get back to it and see how it might be able to help those with GBM.

Such an irony that this stupid GBM struck me after I had started to think about the application of nanotechnology to fight brain tumors.

When I win my battle and become cancer-free, I'll raise billions and billions to buy all old growth redwood trees and save them from being cut down by lumber companies.

I'll use the money to save sea turtles, dolphins, and whales. I'll help clean up the thousands of tons of plastic trash floating in the oceans, concentrated in the huge swirl of water in the middle of the Pacific. I've thought about how to collect the floating trash from an ocean area larger than Germany and France combined. With technology, it can be done and it should be done.

Saya worries that I don't get enough sleep. I'm ruining myself staying awake all night. Papa is urging me to agree that we all fly back to San Francisco. They have already packed the suitcases. They drag me into the car to drive to the Kona Airport. On the way, I delay them by requesting to buy bottles of Hawaiian spring water at a special store. At the airport, I refuse to get into the wheelchair provided by the airline, though I can hardly stand by myself. Papa has to return the rental car. Saya asks a porter to bring our luggage to the check-in counter. I lose my wallet—with my ID, my driver's license, insurance card, credit card, and some cash. Saya runs and runs throughout the airport building, returning to every corner where we have been, and finally finds it, but by then we have missed the plane.

Eight hours later, we are again at the airport. This time we make it. Saya standing and me sitting on a wheelchair, we are lifted to a side door high up in the fuselage, normally used for loading the food onto the plane. It's going to be an overnight flight, February fifteenth, 2010.

Once safely on the plane, I check the latest news on my cell phone. I find out that the plane we missed in the afternoon had an unusually bumpy flight all the way over the Pacific, with heavy air turbulence and surely very unhappy passengers.

"How good that we weren't on that earlier plane," I say to comfort Saya who is almost dissolving in worries.

VACCINE TRIAL

Early morning in San Francisco, still very dark, Saya and Papa check me straight from the airport into the emergency room at UCSF. The medical staff has a quick look at me and then goes about giving me a treatment for my tumor.

I am in much worse shape than I thought. Had I stayed in Hawaii, the doctors at UCSF tell me, I would by now be either heavily handicapped or dead.

The radiologist shows me on the computer screen the MRI, which he just recorded. He was also able to pull up older MRI scans both from Columbia and Hawaii. According to him, while in Hawaii, I had indeed significantly slowed down the tumor volume growth rate.

"Which drugs did you take?" he asked.

"Decadron and keppra."

"These are not anti-cancer drugs."

"But that's all I took."

"In this case, it's hard to believe," the radiologist said. "A GBM doesn't slow down. It keeps growing."

I explained to him that I had my own path—meditation, herbs, very low-carb diet.

"Amazing," was his short answer.

My method was working, but not fast enough. The GBM cells outran me.

However, I also learned from the radiologist that, while focusing on the volume growth of my GBM, I had missed the growth in one direction. This protrusion, which can only be seen by rotating the MRI and finding one particular orientation, must have been the reason why I had this surge in edema and all the other complications.

At the same time, after I had gotten rid of the edema through Dr. DeSalvo's intravenous decadron infusion, I did not feel sick anymore. My mind became lucid again and I was only thinking of how much I would like to do after I get rid of this annoying GBM. So much, so much, so much to do.

———

"I am Dr. Andrew Parsa, your neurosurgeon, and I understand you want your tumor removed."

"That is correct."

"I also heard you are interested in joining our clinical vaccine trial?"

"Correct."

"I understand you have undergone a resection of your tumor already at Columbia, but have never had radiation and chemotherapy."

"Correct."

"In that case you qualify. It is a prerequisite for the participation in our clinical vaccine trial that you have never received radiation nor chemo."

"I have never received radiation nor chemo."

"Are you willing now to undergo radiation and chemo after the surgery in order to participate in our clinical vaccine trial?"

"Yes."

"So, let me get this all straight: you will undergo surgery, then you are willing to undergo radiation and chemotherapy to be enrolled in the clinical vaccine trial?"

"That is correct."

"So, now we need to find a date for your surgery."

Dr. Parsa goes on to explain that tumor cells march out of the boundary of the tumor. Every day, every hour, more tumor cells migrate into the healthy brain tissue.

"Let's do it tomorrow, then," I say, loud.

Dr. Parsa nods but tells me to think it over until seven A.M. the next morning. He takes his leave.

Saya, who has been sitting at my bedside, opposite Dr. Parsa, tells me how surprised she is to see me so eager to go ahead with the surgery. No hesitation this time compared, to the days of agony and torment in Columbia, New York.

I'm surprised myself by how willing I am to accept the treatment Dr. Parsa is offering to get me into the vaccine trial. I adamantly opposed radiation and chemotherapy in Columbia, knowing that they destroy healthy brain tissues and bring all sorts of side effects to the entire body.

Now I definitely want to participate in the vaccine trial. As part of my search for treatment options, I had read about it already in Hawaii. I know, it's the heat shock vaccine, and it looks promising. What this trial does is create a vaccine from my own cancer cells extracted during surgery. In essence, what they will be doing is to treat my cancer cells in such a way that they are still alive but have lost their ghastly ability to rapidly multiply and spread. Those deactivated cancer cells will be re-injected under my skin once every month so that the immune soldiers of my body learn how to recognize GBM cells.

The additional condition to participate in the clinical trial, namely six weeks of radiation for five days a week and long-term chemotherapy are both designed to weaken, with each vaccine shot, the cancer cells that still remain. The radiation treatment is brutal, two Gy every session with three to six MeV hard X-rays and a total dose of about 60 Gy.

It takes several weeks to produce the vaccine. In the meantime, in order to prevent the remaining cancer cells from proliferating as they did so soon after my Columbia surgery, I have to do as much as possible to stop them. At Columbia, I was right to refuse radiation and chemo because that was all they could offer. I was stubborn then, but now there is a new vaccine, made from my own tumor cells after they have been heat-shocked. The vaccine is giving my body an army of well-trained immune cells to fight the tumor.

In Hawaii, I did everything I could to strengthen my immune cells, so that they would be able to attack the tumor. I tried to destroy angiogenesis that feed the tumor. A combination of all the drugs, supplements, meditation, the sun, the ocean breeze, and a good diet started to show the desired effects, but too slowly, snail-like, no real match to the aggressively marching cancerous glio cells.

On February seventeenth, my second tumor was successfully removed at UCSF Medical Center during a two-and-a-half-hour operation. A second surgery is always more delicate and dangerous than the first, but fortunately Dr. Parsa had honed his skill during his residency with Dr. Jeff Bruce at Columbia, my first surgeon. What a wonderful coincidence.

I was aware that Dr. Parsa would take out a piece of my skull, flip back the dura mater below, thus exposing my brain tissue. Then he would place an electrode array to map an area of the brain where he could minimize the damage during the incision and thus find the best area to proceed. He would then cut through the outer layer of my brain, including the gray matter, to get to my tumor. At the end he would place the piece of my skull back and hold it there with tiny titanium screws. In the process, he would have to save as much of the tumor tissue as possible for the vaccine production.

INTENSIVE CARE

Is it night or day? I am lying in a bed, hooked up to a lot of monitoring equipment. Must be post-op. This is definitely the Intensive Care Unit, ICU. I start to look around. I am in a big room, enclosed on three sides by walls, like a bay. The fourth side has a long, cheerfully patterned curtain covering half of a wide sliding door. Muffled voices are coming from behind that curtain.

What time is it? I can't tell, but somewhere on one wall I finally can make out a clock face seemingly stuck at noon. Is the dial not moving? Okay, let's spend some time watching.... It is moving, very reassuring.

I start to recognize more sounds: rhythmic high-pitched beeping behind me, and more than just voices coming through the curtain. What is behind it? I am intrigued.

I remember the surgery started about five-thirty P.M. on February seventeenth. So, unless there was a complication, it must have ended some time before nine P.M. If the clock shows twelve, it could be midnight, or it could be midday. I am not tired.

I'm trying to call. Nothing...

Where is my voice? Maybe I still have a breathing tube stuck in my throat? But nothing's there.

I start with monosyllabic sounds, imagining that I'm a sheep: "*Blaahh.*" It goes on for a while,... Just a monologue of sounds, syllables.... *Eeeh, iiih. Oooh, uuuh.* Soon the alphabet. I try to pronounce it in German, French, English, and then try Japanese *a...i...u...e...o...ka...ki...ku...ke... ko....*

Soon I find words. Words? What do I remember? Saya's voice, her Japanese lullaby songs, childhood songs, more of them—slowly, unsteady. I am remembering Amelia's melodic Swiss German intonation. The sounds evoke an image of the Swiss Alps, snow-covered, in bright sunshine. I was on a train from Interlaken to Zürich, returning from the high-temperature superconductor conference in 1987, with an impossibly clear blue sky, the central Swiss massif of the Jungfrau, Eiger, and Mönch in the Canton Bern, covered in blinding white.

I work myself into a steady stream of words. More and more, I go on. After a while, I get bored of German. Let's try some English. I go on, first tentatively, now with more confidence. I search around the walls for things to read, finding instruction sheets attached to them. Pretty far and small print. Reading them is hard. Intonation is hard. Time is moving. So are my words. I start to talk more to myself, more words. I try to find more sureness. I try to give my voice more variation. I play around with the voice volume, trying to increase, then decrease, modulate it more. The grogginess starts to wear off.

However, I have a question: Where the hell is my breakfast, if this is post-op? Where for that matter are the nurses? I start to feel around my body EKG wires, blinking lights, IV, arterial line, beeping sounds—the reassuring monotony of it all.... The catheter is still stuck in there.

As the clock moves on past one, the curtain is slowly pulled back just a bit, and a smiling face appears, a smile.

"You are at UCSF neurosurgery ICU, and it is February 18, 2010," the smile tells me.

"Where is my breakfast?"

"It's one o'clock in the morning," the smile responds.

I study the smile at the curtain. As it steps into my room, it attaches itself to a nurse, and she steps forward to examine me. I do not recall much about her, but she is very nice. Soon she is at my bedside, and then as she is almost hovering over me, her feminine smells hit me. For the first time I am sure I am alive. She checks my vitals, and asks me to go back to sleep, then disappears behind the curtain.

I watch the curtain, swing back and forth as the air moves in and out of my room, back and forth, back and forth.... The "one o'clock" somehow sticks in my mind, and I keep my eyes closed; and soon I fall into some slumber. I try sleeping for a while, but curiosity about how it feels to be alive just swells over me, so I open my eyes, and soon sounds come pouring out of my mouth again. I become a gushing geyser. I try to recite the U.S. Constitution, which I admire, but I get stuck. I start to recite instead the end of Hemingway's *The Old Man and the Sea*. Most of what I remember in German are Saya's novels. Then I reach back further, and memories of decades past return, and now the time really flies with a vengeance. I recite Voltaire's great poems, but mostly half a dozen or so poems by Arthur Rimbaud, one of the great French poets, who tragically died from cancer at a young age. His poems "Le Mal," "Le Dormeur du Val," and "Sensation" remain forever etched in my memory, and with them the memories of their unspeakable beauty.

Trying to describe the melodic beauty of the French language to most German or English speakers is like trying to explain the melodic beauty of Mozart's "Eine Kleine Nachtmusik" to someone who has grown up on a diet of hip-hop and heavy metal.

Soon I enter a twilight zone between sleep and wakefulness.

———

Until I was thirteen, formal schooling was incredibly boring. I was more interested in reading non-stop, in honing my skills by building intricate model airplanes, in playing piano, and in going to classical music concerts.

Then Mr. Bartenstein took over the stewardship of our class. He was in his mid-fifties, a connoisseur of French literature, history, cuisine, wine, and culture. He was a great teacher with a striking precision of Parisian French. I was too young to appreciate the depth of Bartenstein's knowledge, but later I understood what I had jotted in the margins of the books he had assigned like *Le Rouge et le Noir, Les Miserables, Madame Bovary, L'Etranger,* and *La Peste.* I went on to read everything I could.

Spring 1981 brought graduation and preparation for the ETH entrance exam in Zürich. Around the same time, for some reason, Bartenstein's mind started to wander away. Somebody must have turned him in, because one day Mr. Bartenstein was put into an asylum. I had just seen the movie *One Flew over the Cuckoo's Nest,* in which a completely sane man is turned insane by the treatment he witnesses and experiences in the mental asylum. I became worried about Bartenstein.

With my parents' encouragement during the summer of 1981, I went every afternoon to the mental ward and took Bartenstein out to spend hours with him going through the green parks that encircle Cologne. He was sunk into his wheelchair, his face fallen, and he appeared to be in constant fear, peering around with suspicion. Once in a while we would rest, his frame shrunken, with the sun on his face. I would talk to him and ask him questions. He was silent.

Years later, during a warm autumn day, Saya and I visited Bartenstein in his apartment. He was confined to a tiny room under the roof of one of those huge pre-nineteenth-century buildings with fifteen-foot-high-ceilings and a central wooden staircase. Dirty clothes and dirty dishes lay all over the apartment, and if it were not for an open window and a beam of

sunshine that found its way in, despair would have completely taken over the place. There he was, unshaven and shrunk,
his eyes deep-set. This was all that remained of that giant of a
teacher. Around him rested the remnants of his library, which
once contained thousands of books. It was heart-breaking to
see how embarrassed he was by our witnessing his condition.

Then a bit of small talk with Saya. I was utterly astonished: Bartenstein remembered the time when he was in the
psychiatric ward and I visited him. He remembered that I
came every day. He even remembered the questions I asked
during our walks through the park, and he profoundly
thanked me for having taken care of him. His description
of the details of our afternoons in the park made it certain
that, despite his miserable appearance, his brain had been
working perfectly and had recorded my questions in great
detail.

I deeply regret that I didn't spend more time with him on
that autumn afternoon. Still today I want to ask him about
his childhood, growing up in the border region between Germany and France, next to a concentration camp set up by the
Nazis during the war. He had told the story in class about
how he once looked too closely at what was going on inside
a camp and was caught by the Nazi guards. Somehow they
let him go.

Besides the tremendous influence on my appreciation of
language and literature, Bartenstein had deeply influenced
my personal life. He validated my identity, telling me how
wonderful it was that my mother was Japanese. Racism and
school bullying against Orientals—not only by students but
also by teachers—was the norm in the Germany of the 1970s.

The clock has moved to almost two-thirty. I am filled with joy
and gratitude for being alive. I start to recite Robert Frost's
lines from "Meeting and Passing."

Smile comes back and asks me to tone down the level of my voice.

"I have never been so happy and grateful to be alive," I tell her. "This was my second GBM resection in four months. I can still think. I can still remember, and I can still pronounce words. Isn't this wonderful?"

She bends over me, hugs me, and whispers, "Please just keep it down."

Slowly as the waking hours of February eighteenth creep up, I continue my monologue. After a while, I ask for breakfast. Nope, all I get is a cup of applesauce. I ask for more. I get a second one, then a third. Before long I clean out the station's refrigerator. I should have told Saya to leave a lot of sushi for me.

As I babble on, I also intently listening to the beeps behind me, as well as to the cacophony of voices and machinery of a waking hospital. I hear steps right outside my bay and the curtain pulls back. Dr. Parsa comes in.

"Look at you!" he exclaims with excitement, his hand pointing my way, his face lighting up with a broad smile, his eyes sparkling with utter delight. He steps forward.

"Thank you, Andy. How did the operation go?"

Parsa explains that there was a section of the tumor, three to five percent of the whole, that he had to leave in, because taking it out would have seriously risked a permanent damage to my brain. He checks my fingers and toes, to be sure they are properly moving. "Fantastic, Mino, you're doing so well, you can go home on Saturday."

I'm elated. I can go home, I can go home, after six months, *go home*!

"And you still want to participate in the clinical vaccine trial?" he asks.

"Of course, and whatever it takes to get in."

His smile broadens even more.

GLOOMY LIGHT REHAB CENTER

I have been moved from the Intensive Care to the Acute Care Unit of neurosurgery at UCSF. The room is smaller than at Columbia, but the nursing staff is far superior. The nurses here are not only better trained but also more compassionate. Saya doesn't have to stay with me during the nights, because the night nurses seem to know where an oxygen inhaler hangs and what they have to do in case I need a wisp of extra oxygen.

One after the other, my friends and colleagues from NASA Ames come to visit me. Pete Worden, our Center Director, and Al Weston, Projects Director, had to wait in front of my room almost for an hour, because I was in the middle of physical and occupational therapy, followed by being washed and changed. After days of neglect I did not want to appear unshaven in front of Pete and Al.

Everybody is happy to see me, and I'm grateful to see them. I'm happy to be able to speak, laugh, and recount what has happened and what is happening before my eyes. My NASA friends are reporting to me about our projects as if I had never left them. Deb Feng brings a lot of gyoza and Hege al Ali is intent on spoiling me with her food. Both are excellent cooks, giving me the joy to eat more. I now have to learn how to stand up without support, to stand on my feet, and to walk. I have to learn how to use my left hand, which is still partly paralyzed, to hold a pen or a cup without dropping them.

That's the only damage left from the surgery. At Columbia, after my first surgery, I also had to train my left hand and left foot, so that they would regain their normal functions. At that time, I had achieved this in three days and I was able to drive home by myself after the discharge. This time Andy Parsa, delighted to find me doing so well at ICU, told me I could go home this Saturday.

I'm looking forward to going home in two days. Home, after six long months and a long journey through Pittsburgh, New York, and Hawaii.

Last night I heard cats—many cats—caterwauling somewhere next door. The wailing sound was so mournful and sometimes so crude that I could hardly bear it. After a while, I called the night nurse, who explained to me in a low voice that there were no cats in the acute care unit, but there was a woman patient next door with a recurrent GBM tumor who had lost her speaking ability. She was desperately trying to say something but could not form words.

I felt a chill on my back. I must get healthy as soon as possible and get back to the nanotechnology neurology project I had started. I must be able to find a solution, applying nanomaterials, to reach damaged regions in the brain. I'll be able to help this woman and many other GBM patients, if only I can start my nanotechnology project. I began drawing up in my mind the details of such a project and how to approach it. I fell asleep.

In the morning, the chief of the Acute Care tells me that, after such major surgery, I must stay for ten more days, or at least for one week, for observation because unforeseen complications might occur. He also says I have to stay for the physical and occupational therapy. I'm upset and tell him that, according to Dr. Parsa, I can go home this Saturday—in two days. The chief shakes his head, insisting that Parsa is not in the position to decide when I'll be ready for discharge from Acute Care.

What a shock. Only a few minutes ago I was exuberant, thinking of going home after six months and thinking of my NASA team with whom I would restart my projects one by one. I was humming with joy.

No more.

My discussion with the chief is getting volatile. Saya runs

out of the room looking for Dr. Parsa. She goes around everywhere asking nurses for his whereabouts. She knows that I
spontaneously liked Dr. Parsa when he walked into my life.
I liked his obvious competence, but also his French flair and
slight accent. I trust him. When he says that I can go home,
I go home. Nobody can change my mind but Parsa himself.
Saya knows that.

She finally locates him and asks him to come to see me,
even only for a minute. But Dr. Parsa doesn't remember what
he told me at the ICU in the early morning hours after my
surgery. Anyhow, he is busy and has no time for me because
he has new patients to look after.

Meanwhile I get mad at the chief, who repeats the same
stuff no matter what I tell him. I start banging his chest.
A nurse comes and sedates me with some strong injection.
Instantly I fall asleep. Next morning the Chief agrees to discharge me to a rehabilitation clinic.

I tell him again that Dr. Andy Parsa assured me I could
go home—because I'm doing so well. I don't want to go to
some rehabilitation clinic. I want to go home.

Nobody in Acute Care listens to me. Everything has
already been arranged, and the rehabilitation clinic is waiting for my arrival.

The Bright Light Rehabilitation Center turns out to be a
living nightmare, if I have ever seen one. Being sent here is
the worst thing that has ever happened to me emotionally.
The staff at UCSF told me that in the Bright Light Center, I
would meet well-trained nurses and would receive inspiring
physical and occupational therapies all day.

In reality, except for one and a half hours of the physical
therapy in the morning, the so-called "Bright Light" forces
me to lie in bed, idle all day. I want to exercise more, exercise at least for a few more hours every day, but this is not
allowed. Physical therapy is available only for one and a

half hours, but not for more. The patients I'm seeing are all decrepit and infirm, mostly stroke victims, unable to utter a word. Through the open door of my room I see once in a while a bent-over silhouette pushing a walker along a gloomy and soundless corridor. Somewhere, in a neighboring room, someone is snoring or goes into a long fit of coughing and spitting out phlegm. In the dimly lit rooms, as far as I can determine, the patients are lying in their beds without visible sign of life. It's eerie. I'm utterly confused.

I begin to worry that I'm losing time to reacquire my ability to walk and to hold things with my left hand. If this continues, I'll be permanently handicapped and unable to return to work at NASA on any nano-neuroscience project. I want to go outside, sit and take some steps in the sun, breathe the spring air, and look up to the sky. But none of this is allowed. Unfortunately, I'm not yet able to stand up by myself, let alone walk a few steps. I need someone to support me, but nobody answers my bell. More than once at night, while trying to climb out of my high hospital bed, I have fallen on the hard floor. When that happens, I lie there for several hours, because nobody hears my calls.

In time I start to suffer from a mental freeze. I can follow everything around me with the movement of my eyes, and I can hear everything, but I can't talk anymore. The staff is clueless on how to handle me.

Worst of all, Saya can't come to see me at the Gloomy Light Rehab because she became very sick right after UCSF. She has high fever, headaches, and a severe sore throat; and she has started to take antibiotics. If she weren't sick herself, she would have immediately grasped the situation and would have rescued me. She would be the person I could tell how much I was longing for the bright sun and the vast sky. She would take me out of this Gloomy Light Rehab. We could laugh about a short story by Gabriel Garcia Marquez, "I

Only Came to Use the Phone," a macabre tale about an acci-
dental tourist, a completely normal woman, who by mistake
ends up in a mental hospital, being treated as deranged, and
gradually loses herself.

Papa visits me often, but he talks only with some staff
members at the reception. They must have convinced him
that I'm doing fine and that my ability to stand up and walk
is improving rapidly. Papa seems always to be in a hurry,
never stays for longer than a few minutes, and disappears
again swiftly to return to his own work.

Hege al Ali comes to my rescue. She listens to my ICU
story and encourages me never to give up. She brings delicious home-made food, telling me about her little daughter
and her experiences in Qatar. She has kept me sane during the
ten long days in this torture chamber, which I call Gloomy
Light, because—for me—nothing bright is coming out
of it.

The Gloomy Light people still try to keep me longer, but
I'm doing everything I can think of to get out.

On March 3, 2010, Dr. Mathew Goodman, discipline
scientist and program manager at DARPA, wrote to Papa:
"I spoke to Mino by phone this morning. He was alert and
lucid...I must say I stand in awe of his amazing intelligence,
focus, stamina, and desire to help change the world for the
better. I hope his treatment will bring complete remission and
a speedy recovery."

March 3, 2010 was also the last day of my living hell at
the Gloomy Light.

SLEEPING PHASE

I am finally at home, though I can't walk and can't use my left
hand. Saya has recovered from her flu. She takes charge of
scheduling visits to doctors, therapists, and practitioners. She
cooks, does laundry, and cleans the house; she also supervises

a night nurse for me and a driver who helps me get around from one appointment to the next.

Oh, yes, driving. The DMV yanks your driver's license once you have a major brain surgery. Initially, I insisted on driving myself as I had done in New York after my Columbia surgery, but that's not allowed in California.

For the next few weeks, a typical day would look like this: physical or occupational therapy the whole morning; then radiation at Stanford Medical School, which takes about one and a half hours including driving, radiation for five to seven minutes of actual beam time. Superb technical staff that treats me as if I were being fed caviar. So far, no nausea or fatigue. In addition, my driver brings me to my Traditional Chinese Medicine (TCM) treatment three times a week for two-hour sessions. The TCM practitioner, Fred Dong, a short, jolly sixty-year-old Chinese, gives me acupuncture on my whole body, including my head. Extremely soothing. He gives me a week's worth of herbs, which Saya boils for two hours to extract the essence. The TCM treatment is designed to counterbalance the damage done by chemo and radiation. Every night I must take a chemo capsule with a tall glass of water on an empty stomach before going to bed. Once a week I see my local oncologist, Dr. Shane Dormady.

I'm improving visibly day by day. A week ago, I could hardly manage to walk a few hundred feet in Questa Park. On April third, however, to the sheer joy and amazement of my parents, I managed to walk the full one-mile loop in the Henry Cowell Redwood State Park among the 250-foot-tall redwoods.

This Redwood Park was the place I showed my parents for the first time in 2006. They were stunned seeing these magnificent ancient trees. Well over a thousand years old. We walked the loop in silence, in a feeling of awe. Since that first visit, I have often taken them to the redwood trees, not only in the Henry Cowell State Park but also in many other parks in Santa Cruz mountains, in particular to Big Basin, my spiritual refuge.

There we hiked all day, or at least half a day, and I was always ahead of them on the rugged, steep terrain or on narrow uphill–downhill trails. Now my parents stroll by my side, adjusting their speed to my slow steps.

The past six weeks have been taken up by chemo, radiation, occupational and physical therapy, acupuncture, herbs, and doctor visits. Now I realize that all this has taken an enormous toll on my strength. I find myself in a sleeping phase, requiring fourteen to eighteen hours of sleep every day. I am hibernating in my bed, as if I were a bear in a den. No sound disturbs me. Even emergency sirens of fire engines and the weekly garbage truck don't interrupt my slumber.

In the beginning, Saya was worried, thinking that something had gone wrong with me because nobody had warned us what to expect after six weeks of radiation, 2 Gy with 3–6 MeV hard X-rays, altogether 60 Gy. The radiation was designed to focus on the remaining cancer cells, but in order to reach their target, they have to penetrate skin, skull, blood vessels, and healthy brain tissue. Of course, such intense radiation does non-trivial damage to all the healthy cells in its path and to the entire immune system. So far, I haven't been able to find any concrete information about the extent of the damage inflicted on my body, both physical and physiological. I decide not to worry about it because I simply have to go through this radiation and the simultaneous high-powered chemo, Temodar, in order to qualify for the vaccine trial.

In time Saya understood that, by hibernating in my den like a bear, I'm recovering from the stress of the combined radiation and chemo treatment. No fever, no headache, no nausea, and I'm eating everything she gives me with great joy during the short hours I'm awake.

Once I was dreaming, seeing hundreds of dolphins in linear procession, performing acrobatic jumps. I even saw their happy faces and smiling eyes. It must have been in Honaunau, because vast black lava flow spread before me, and the ocean beyond with the line of jumping dolphins. The sky was the same blue as the ocean. The crests of the waves were white. As the dolphins swam out of sight, I found myself still sitting on the gray sand beach. A large green sea turtle was taking a nap nearby. I shooed away noisy tourists.

Another time I dreamed I was visiting the National Gallery in D.C. for an exhibition of drawings and etchings by Brueghel, Holbein the Elder and Holbein the Younger, Dürer, and others. To my utter surprise, I saw my own drawings among the fifteenth- and sixteenth-century masters—my interior of Notre Dame de Paris, my cityscape of Cologne across the Rhine, my silhouette of the Grossmünster in Zürich, and my sketch of the Kenkun Shrine in Kyoto. I called Saya. In the next moment, I woke up.

I own three large prints by Norman Rockwell, including his "Triple Self-Portrait," which I bought at an exhibition and had framed. Whenever I look at them they make me smile. Rockwell's "Gossip" is just hilarious, but it was too expensive for me at that time. I didn't buy it. Once I saw an exhibition of Edward Hopper, so powerful that it almost knocked me down. Beauty of solitude, I thought. However, I would never hang any of his work in my room because everything looks so desolate, unconnected, frozen in strong, hard contours, colors and shadows—buildings, people, even animals. Both Rockwell and Hopper lived in New York around the same time and were of about the same age, but they observed and presented distinctly separate realities, as though they were living in countries far apart.

A few nights later, during another dream, I saw Canada geese. I remember them from Goddard Space Flight Center in Maryland. They were all over the place. Magnificent birds, with brown-beige earth-tone bodies, black heads, tails and wing tips, and white stripes on their long necks and on the sides. They graze in the wide-open green fields and often cross the roads with necks stretched, making sure that car drivers see them. I dreamt they were breeding. I was looking for the pair that had built its nest right next to the entrance of our building. Somehow, I couldn't find the nest, and I woke up.

The owl in the animal refuge of the Glen Helen Nature Preserve in Ohio appeared in another dream. The owl couldn't fly because of its broken wing. When people approached the enclosure, it would hide behind the leaves. I stood there, talking gently to the owl. It glanced at me from a distance and started to walk sideways toward me, hesitantly. I sang Saya's lullaby. The owl came to the branch right in front of me, and I touched it through the bars. I told it that I would come back again soon.

A week later, visitors were surrounding the enclosure, but the owl refused to come out from hiding. I waited until the visitors moved on. Then I approached the owl, talking to it and singing as I had done in the first time. After a short while the owl came out, walking sideways step by step toward me, bringing a dead mouse as a present. It didn't move until I accepted its gift. When I took the mouse from its beak, saying, "Thank you," the owl ruffled its feathers and flapped its healthy wing. In my dream this owl flew towards me and landed on my arm. As I sang Saya's lullaby, it flapped both its wings to show me that it was healthy again. So real was the vision that, when I woke up, I looked all over my room for my owl.

My sleeping phase lasted nearly two weeks. At its end I was given my first heat-shock vaccine injection at UCSF.

Antigenics Inc, the vaccine manufacturer, calls it HSPPC-96. They have produced enough doses to keep me in this trial for over two years. Dr. Parsa writes on his website: "We generated some remarkable anecdotal efficacy data as well as compelling immuno-monitoring data." Antigenics is a bit more specific: "The first 20 phase 2 recurrent glioma patients treated with oncophage showed a median survival of 10.1 months, with at least six patients surviving 12 months or longer. These early clinical data appear to compare favorably with the long-established historical median survival of 6.5 months, and with the recently reported median survival of 9.2 months with Avastin in patients with recurrent high-grade glioma."

By now I have survived eight months since the diagnosis of my GBM tumor at the end of August 2009. Joining the heat shock vaccine trial means that the next phase of my battle begins. I will continue physical and occupational therapies. I will continue acupuncture and broader TCM therapy, which has already shown impressive results. Saya is busy scheduling doctors, therapists, and my driver; cooking dinners for my friends from NASA; and paying bill after bill. Papa puts handrails all over our house. We stop having a night nurse, because I can walk again, though only with a cane.

FLASH OF MEMORY

Between appointments, I sit at my computer typing with my right hand. I continue my blog, which by now has attracted more than forty thousand readers. I write emails. I read scientific papers and scientific news. On one of those days, I read that Andrew Lange has committed suicide in a Pasadena hotel.

Memories flash back to my time as a post-doc with Dr. Andrew Lange, an ambitious young assistant professor at the University of California Berkeley. He gave me the task of completing an ultra-low temperature cooler, the core of the FIRP instrument on the joint U.S.–Japan satellite mission,

IRTS (Infrared Telescope in Space). IRTS was expected to tally up stars and galaxies and to peer back in time as far as possible, billions of years. Andrew Lange was in charge of the FIRP, one of two instruments on this mission to be supplied by the American side.

I determined that Andrew Lange's original design of the FIRP, which had been put together by a previous post-doc, did not achieve its target subKelvin temperature. I determined that the design was flawed. Only eighteen months left until launch. Not enough time to redesign and rebuild such a complex, centrally important device, everybody said, and Andrew Lange freaked out. A brilliant astrophysicist, he was expected to climb to the top of the cosmology field. Worldtop. He panicked and put all blame on me for the malfunction. The only thing he could do was to shout in my face: "Fix it! Tweak it! Stupid!"

I redesigned the FIRP, rebuilt it from scratch and reached at the first try 260 milliKelvin—the temperature needed to record the far infrared radiation coming from near the edge of the universe. I tested my FIRP at NASA Ames. I put it through grueling high vibrational loads, mimicking those during launch, and subjected it to extreme temperature cycles. It passed all tests, and everything was completed on time. The U.S.–Japan joint IRTS mission launched on schedule. It was a highly successful mission, generating a wealth of data, which increased our understanding of the universe. The IRTS became the only satellite in the history of space science that was recaptured by the Space Shuttle after completion of the mission and brought back to Earth.

My redesigned FIRP, one of the most successful instruments on the IRTS mission, achieved a world record, the lowest temperature for the longest time in space, and it took more than a decade for the record to be broken.

Andrew Lange declared the FIRP, the crystallization of

my expertise in design and instrument building, to be his creation, and his creation only. He flew to Japan to retrieve the FIRP from the IRTS spacecraft before it was put on display in the Museum of Natural History in Ueno Park, Tokyo. He took the FIRP with him to the California Institute of Technology in Pasadena, where he had become professor and would eventually be nominated Dean of Physics, Mathematics, and Astronomy.

Thinking back to my time at Berkeley I remember Harry Bingham, Professor Emeritus. Upon his retirement, he had transferred his entire lab, more than 2,000 square feet, to Andrew Lange—an invaluable asset for Andrew, who had just started his career and needed room to expand. I spent many hours talking to Harry Bingham. He had a wonderful sense of humor and was always cheerful, hiding the pain of his cancer behind his smile. Two days before his death I still chatted with him. Andrew Lange was away, traveling. When he returned, I was eager to tell him about Harry's passing. I went to his office: "Andrew, I have something very sad to tell you. Harry Bingham just passed away."

Without looking up from his desk, Andrew snapped, "What else is new?"

I'll never forget that moment. I started to better understand Andrew's way of looking at people. His ambition was boundless, driven by a fierce energy from deep within, a dark energy that made him emotionally so empty.

On January 21, 2010, Andrew Lange committed suicide by hanging himself in a hotel room in Pasadena. In the end, he directed his interminable dark energy against himself.

WORKING HARD AGAIN

Half Moon Bay is the closest beach from our house. The windy sand dunes along the Pacific Ocean always seem shifting. We go there sometimes in the late afternoon and I practice walking on the sand. I can stroll with bare feet, and it doesn't matter if I lose balance and fall down.

Papa and Saya stay by my side just like long, long ago, when I first began toddling on a trail in a forest outside Göttingen as a one-year-old with my hands in their hands. Someone took a picture of us from the back, showing how my parents were bending over toward me and how I was stretching my arms up toward them. According to Papa, I liked my first walk so much that I toddled and waddled the whole afternoon until I fell asleep while still walking. The photo was exhibited in a gallery with the title, "First Stroll." The photographer gave us a copy and we have it in our album.

At low tide, the beach spreads out wide with wet and supple sands where I feel with my bare feet just right amount of resistance. A large pile of seaweed, many different shells, sand-covered logs, and colored stones are scattered around, and small birds with long legs dart around, picking their dinner out of the sand. Saya sings her favorite songs, and I start with Papa our discussion on our next publication. From my student years in Zürich until now, we have published more than eighteen papers together and we have still a long list to go. Everything moves in its course, as if nothing dreadful has ever happened to us in our life.

On a modulating dune higher on the beach, we sit down and wait for the sunset. The sky changes its color from light blue to pale rose and slowly to burning orange. I am overwhelmed by the silence before the sundown. We hold our breath and the sound of waves becomes limpid.

A swarm of blue-white birds flies in formation, changing from a straight line to a circular, then a triangle, as if they've been choreographed and are now rehearsing for a stage performance. We are the sole spectators.

The stars begin to shine, and a frosty wind reminds us of home. On our way back to the car Saya and I sing together

"Yuyake koyake," one of children's songs we have sung countless times. Here I give my English translation:

> *The sky is blue, turning orange and red*
> *A gong echoes from a far-away temple*
> *Let's fly home together*
> *Together with the ravens.*
>
> *At home we'll look again at the sky*
> *Watch the large round moon rising,*
> *As the birds dream in their sleep*
> *The sky is filling with stars.*

After my first vaccine shot, three weekly booster shots follow. I become very hungry. I go through an eating phase, during which my base metabolic rate increases—eight sandwiches for breakfast, followed by two whole pizzas for lunch, sixteen-inch, with mushrooms, zucchini, anchovies, and spinach. Dinner is salmon and roast beef accompanied by a huge bowl of seaweed salad.

This reminds me of my early student years in Zürich, when I devoured lunch at the university cafeteria—spaghetti Bolognese and beef stew or chicken risotto, going back with my plate for second and third helpings. The women behind the counter served with a smile or sometimes shook their heads in disbelief. In Japan during intensive data analysis of our satellite mission, IRTS, I could eat one ramen, one large tempura dish, fifteen sushi, fried tofu, and parboiled and pressed spinach, one after the other, for lunch.

During my eating phase, while getting the vaccine booster shots and the weeks that follow, I have tremendous energy. I've gradually regained control of my left hand. I'm able to use it again for typing. Thorough examinations at Stanford Ophthalmology show that the inner layer of my eyes, including the retina and the blood supply, are unscathed by the brain tumor. I have maintained full visual range. I should soon be able to walk without a cane. All I need to do is train the joints and muscles of my left leg and foot three times weekly by physical therapy and acupuncture.

My schedule is full, with MRI, blood tests, ultrasound imaging, plus TCM, PT, and OT, all at different hospitals and clinics. Also, I have to present myself every other month to Dr. Shane Dormady in El Camino Hospital Oncology and to Dr. Michelle Cooke in UC San Francisco Neuro-oncology. My driver takes me to all these different places.

Once, while I was still quite weak, I went into the men's room at Stanford, but could not get out again. The metal door was so heavy that it could have been built for a solitary confinement cell in a jail. I called, shouted, roared, pounded with my right hand, all to no avail. Finally, after more than half an hour, a middle-aged man opened the door.

That incident motivated Saya to accompany me always and everywhere. She walked next to me when I had to cross busy streets from my car to the doctors' offices; when I had to pass through non-automatic doors or navigate my way through meandering patients or when I tried to board an elevator while somebody in a hurry was trying to push his way in at the last moment. Never before had I noticed such hazards lurking for someone like me, who now has to concentrate on keeping balance. I made it. I never fell down.

Between my hectic appointments, I'm still able to do many other things. Foremost those unipolar pulses Papa told me about. They are short pulses of electromagnetic energy, very intense and short, a fraction of a second long, as if a radio station were switched on and off seven, ten, even twenty to forty miles below the surface of the Earth. We know they must come from rocks that are squeezed by the humongous tectonic forces down there. We know that squeezing rocks produces electric currents. The faster they are squeezed, the more powerful the currents. But why do they come as sudden pulses? On and off within a fraction of a second? Why with only one polarity...unipolar? Something very strange must be going on down there, deep below. We don't know...yet, but we'll surely find it out.

And then I have the TremorSat mission on my agenda, a multi-satellite project I put together with Papa. It's to monitor earthquakes from space before they strike. Papa has spent years finding out the different signals that the Earth sends out before major earthquakes—bewilderingly many, infrared anomalies among them. Nobody seems to know why they pop up before big earthquakes, but Papa and I have a pretty good idea—a quantum mechanical effect when vibrationally excited atoms at the Earth's surface de-excite and emit characteristic infrared light. We spend hours talking about the best way to prove it. I help with the analysis of the spectra that Papa and his coworkers have recorded in the lab. Very distinct spectra. Before major earthquakes the Earth should send out infrared light with the same spectrum, and we should be able to record it from space.

That's why I had set up a working group at NASA Ames to discuss my TremorSat mission—a constellation of nanosats, hundreds of them, with infrared cameras looking down on the Earth and recording the infrared emission at night. Plus, sensors to record perturbations in the ionosphere. That's one of those crazy ideas Papa and I have been cooking up, mostly based on the work we published together just before this stupid brain tumor forced me to take a detour. In this paper we reported on the air over one end of a rock becoming ionized when we stress the other end. Massive ionization, producing positive and negative air ions. We are sure all this has to do with the infrared radiation and with the perturbations in the ionosphere, but we don't yet understand enough of the details.

Between all these enthralling discussions, I'm finding time to come to the NASA Ames Research Center, meet with Pete Worden, and talk with Mike Bicay, the Director of Science, a very knowledgeable astronomer, broadly read and excellent at presenting complex scientific issues, even to a lay audience. Matthew Goodman visits from DARPA. He is considering giving our team much-needed funding. Francis Ghesquiere from the World Bank stops by to discuss additional financing options, if countries in the Third World that are pummeled again and again by big earthquakes would get involved.

I pull in Jason Lohn from Carnegie Mellon University, who can help us with his new antenna design to optimize the data transfer to and from our TremorSat constellation. I'm considering using Patrick Hogan's World Wind program, which is so much superior to Google Earth. Excellent and perfectly suited for our TremorSats.

Claes Hedberg, whom I have never met before, flies in from Sweden to discuss with us Papa's ideas about the mechanical properties of rocks when they are stressed. Strange—rocks are getting softer when we squeeze them. An amazing observation. Papa actually posited this and is so happy to find out that it's actually true. I enjoy every minute of the discussion with Claes. It's a wonderful time.

GRAND MAL SEIZURE

On July seventh, 2010, I'm scheduled for an MRI, my eighteenth since August 24, 2009. I'm also scheduled for my fifth vaccine shot at U.C. San Francisco. The MRI results are as good and reassuring as one can hope, no evidence of any tumor growth. It's the best MRI since right after my first surgery at Columbia.

Dr. Michelle Cooke, a youngish-looking, elegantly dressed

neuro-oncologist, speaks with a gentle voice. She has been assigned to be my neuro-oncologist. She points at my MRI image on her computer screen and tells me that my overall health could not be better.

"You are doing well," she says. "You never had a seizure. I'll discontinue your anti-seizure medication. You don't need it anymore."

She performs routine eye tests to determine my peripheral vision, snapping her fingers. "By the way," she adds as a kind of afterthought, "when you start chemo again, you'll get sleepy. This time the dose is three times higher than before. I'll prescribe Ritalin."

Papa asks her: "What is that...Ritalin?"

"A mild stimulant, routinely given to school children to help them focus their attention."

"Focus attention? What are you talking about? Mino doesn't need any stimulant to focus attention."

"No, no, Ritalin is very mild. Mino will be able to better manage the side effects of the chemo."

From that day in July onward, I faithfully follow Dr. Cooke's instruction to stop anti-seizure medication and instead take Ritalin. Daily. My monthly five-day round of chemo continues.

During the summer friends and colleagues come to visit me between medical appointments. We discuss science in our living room. Sometimes we gossip and laugh quite a lot. Usually Saya invites them to join us for dinner. Chatting, eating, and laughing with my old friends, I feel as if nothing had happened...as if I never had a GBM. My last MRI shows no more tumor. All I need to do now is eradicate the last GBM cells that may still be hiding somewhere in my brain.

The hot days are over. The air has turned crisp and fresh. A group of NASA folks and Papa take me to Big Basin Redwood State Park. We hike a long trail up for hours until I can't move anymore. Alisa Hawkins, who organized the trip, hurries down to the Park Headquarters to look for help. She informs the rangers, who come to pick me up with a truck. Alisa is still there, waiting to make sure I come back safely. Late in the afternoon I return home with Papa and with an enormous appetite. That night I sleep soundly for more than ten hours.

Jacob invites me for dinner. We eat in a Chinese restaurant on El Camino Real. Toward the end, just as the desert comes, the fingers of my left hand tingle. Suddenly I feel strange as never before. I feel my whole body cramping up. Stiffening. Falling backwards with my chair.

I pass out.

Here is what I recall.

I hear a siren blasting. Very loud. Why a siren? I'm lying on a hard surface, a sort of bed, but moving. Outside, behind a window, houses and trees fly by. The face of a young man comes into my view, a man in uniform. I ask him: "What's going on?"

"We're taking you to the hospital."

"Why?"

"You had a seizure."

"A seizure? Impossible," I said. "Nonsense! I've never had a seizure."

"Yes, you just had a seizure, a very bad one, a grand mal seizure," the man replies calmly.

"No, I haven't!" I try to grab his arm. This movement sends a sharp pain down my back, a very sharp pain.

The uniformed man stays calm, asking me questions like what my name is, what day of the week it is, what date. I answer them all.

"Do you remember where you have been?" comes his next question, while he's checking my vitals.

"In the Chinese restaurant. Why do you ask?"

"What were you doing there?"

"Eating, of course."

"What is the last thing you remember?"

"Dessert, a fruit dessert." The image of the dining room comes flooding back, mixed with the siren of the ambulance. "Where are we going?"

"To the hospital. El Camino Hospital."

The ambulance turns a curb, causing a stinging pain that runs through my lower back. I'm screaming.

"We're almost there," the man says with his annoying calm.

The ambulance slows down at the emergency entrance. The back doors open, and my stretcher rolls off the ramp. It makes me scream again in pain.

Then the usual paperwork. Out of my wallet come my ID and the insurance card. The minutes crawl by like molasses. I make a mental note not to forget to retrieve my papers later.

My vitals are checked again, this time by a friendly nurse. They transfer me to a gurney. More questions, such as if I had pain, and, if so, how would I describe it on a scale of one to ten.

"Around ten," is my answer. "In my lower back."

They poke my right arm with a needle. "What are you doing?" I ask the nurse.

"We're giving you anti-seizure medication."

A doctor comes in a pale green work coat. I hope he can do something to relieve my excruciating pain. It's getting unbearable.

"Patient has had a grand mal seizure," reports the nurse. "He says he has pain in the lower back."

"What is your name?" I ask the doctor.

"Dr. Dan Fox, head of the ER."

"I had two surgeries for GBM in the last twelve months, first at Columbia, New York, then at UCSF." I rattle off my clinical history, "...and by the way, Dr. Fox, my primary care oncologist is Dr. Shane Dormady."

"Oh, Shane! Good friend." Dr. Fox turns visibly from being formal to being relaxed. "Oh, you're Shane's patient," he nods approvingly and explains that the nurse has already given me some anti-seizure medications, so that no more seizures would follow now.

"Thank you," I say, enduring the agonizing pain with every slightest movement of my body...even just talking.

"We'll have a CT scan of your head to make sure there is no bleeding. Later an X-ray of your lower back. We'll do everything we can to find out what the massive seizure has done to you."

"Are you allergic to CT contrast material?" asks a different nurse, while hooking me up to about a dozen EKG leads.

"Look, I had about sixteen CT scans in the last twelve months. It's like I'm getting contrast material for breakfast."

"Great." She smiles at my joke. "First, we'll get the CT scan of your head."

At this moment, as my gurney is being moved out of the room, I see Saya. What a surprise. She bends over me and takes my hand, "Minochan!"

I suddenly feel so relaxed. Despite the pain. Then, as the gurney starts moving, my stomach feels like turning over. "Ahh, I must throw up!" I gasp.

"He's throwing up," Saya screams. "Help!"

A tiny pink tray appears in front of my face, pushed against my chin. Too small...that's all I can think, but it's too late. I vomit all my dinner into the tray, which instantly overflows, onto Saya's hair, onto her red sweater, and all over the hospital bed and the floor.

"Minochan, get it all out," I hear Saya say. "Don't worry. Get it all out."

It keeps coming and coming. At the end, I feel unburdened.

The nurse and a helper transfer me to another bed. As I'm being rolled off, I see a janitor starting to clean up the mess.

CT scanners are awfully cool places. I always think of them as a cross between a guillotine and an engine that sounds like the helicopter taking off in the movie *Apocalypse Now*. Very noisy, but providing minutes of solitude. Then back to my room.

"How many seizures has Mino had until now?" Dr. Fox asks Saya.

"None, absolutely none. Mino never had any seizure." Saya tells Dr. Fox that, exactly one year ago, October fifth, 2009, I had my first surgery. Since then I've taken anti-seizure medication, daily, first depakene, then keppra.

"Then I don't understand why a seizure?" Dr. Fox sounds perplexed.

Saya tells him that my neuro-oncologist at UCSF, Dr. Michelle Cooke, removed me from anti-seizure medication in July and, at the same time, prescribed Ritalin.

"Ritalin...Ritalin...*Ritalin*?" Dr. Fox rolls his eyes. "Ritalin is known to cause seizures, particularly after brain surgery...Who is this woman?"

"Dr. Michelle Cooke, neuro-oncologist at UCSF."

"Crazy. How can one stop anti-seizure medication to a brain tumor patient and give Ritalin? It's like rolling out a red carpet for a seizure. Crazy...."

I'm carted off again, this time to the X-ray room. Two quick images of my lower back.

When I return, I see Dr. Fox on the cell phone in an animated conversation. I ask Saya to whom he's talking.

"Rose Lai," she whispers. It turns out Saya had called Rose Lai, my neuro-oncologist at Columbia, though it is two A.M. New York time. I hear Dr. Fox and Rose talking about Ritalin and how unprofessional it is to stop my anti-seizure medication.

My CT brain scan appears on the computer screen. Dr. Fox examines it intensely. "No lesion, no anomaly, no internal bleeding. That's good news."

But my back pain is intolerable. I want to say, *What the hell is going on?*

"Do you have allergies for any pain medications?" a nurse asks me.

"Yes, I am allergic to Percocet."

They give me twenty mg morphine. It works near instantly. Finally relaxed, I'm resting in my bed.

Deb Feng shows up—what a surprise.

Time keeps creeping by and it's already two A.M. Where did the hours go?

Wish Papa were here.

Suddenly, Shane Dormady stands by my bedside. With hundreds of cancer patients, Shane is always up to his ears. Why did he come? At two A.M.?

Shane bends over me.

"Thanks for coming," I say hesitantly.

He smiles.

"You're afraid I wouldn't make it?"

Putting his hand on my shoulder: "Yes," he says.

COLLISION

Coming home from the ER, I sleep for a few hours. As early daylight wakes me up, I feel again the pain in my lower back. Increases hour by hour. Persists through the day and night, into the next day.

A new series of MRI scans at El Camino Hospital shows a gaping lesion in my lower back. The force of the muscle contractions during my grand mal seizure have displaced my L4 vertebra, moved it sideways, off my spine by as much as one centimeter. Both disks above and below L4 have been seriously deformed.

Back at home Saya feeds me a few spoonfuls at bedside. If I eat more, I'll nauseate and feel like throwing up. Everything has to be put on hold—chemotherapy, physical and occupational therapy, and TCM.

Sitting next to my bed, Saya talks on the phone with Michelle Cooke, whose incompetence was the cause of my seizure. I can feel how strenuously Saya is trying to control her voice and remain calm. I hear Dr. Cooke answer in a haughty tone that seizures can happen anytime. "Ritalin is a mild stimulant, routinely given to school kids. Well, if you prefer, I can prescribe for him anti-seizure medication again. That's all right with me."

When Saya talks about my displaced L4 and the deformed disks above and below, which give me excruciating pain, Dr. Cooke comments with her gentle voice: "Sorry about that."

Papa and Saya take me by car to an orthopedist, Dr. Parker, several blocks from our house. I can barely walk, even a few steps. Looking at the MRI images, which we've brought along, Dr. Parker tells us that the severity of my spine injury suggests that I must have taken a very hard fall, probably backward together with my chair. Unfortunately, nothing can be done about this kind of injury. Trying to repair it by surgery can make it worse. Dr. Parker orders total bed rest for at least two weeks. No chemo during this time. After that I should come back for a new MRI, and he would then tell me when I might resume physical therapy. "If necessary, under strong pain killers," he says.

"Eventually, your pain should become tolerable."

"When?"

"In a few months, I hope."

I can't swallow anything, except drink a bit of fruit juice.

At the end of my bed rest a message arrives from Dr. Cooke at UCSF, admonishing me that I have not taken chemo. She orders me to start right now, increasing the dose for the next five days and then come to UCSF for the next injection of my heat shock vaccine.

I quickly count the days in my head. This creates a scary situation.

Before my seizure, I read about the biochemical reactions produced by my chemo, Temodar, a powerful chemotherapeutic agent. I've read about the biochemical and physiological reactions it produces in the body. The extreme opposite of what my heat shock vaccine tries to achieve. The chemo is an immune suppressant. It inhibits cell division, not only of cancer cells, but also of the immune cells. My vaccine is an immune booster. Temodar and my vaccine don't get along together. That is why my protocol requires a hiatus of two weeks between the chemo and vaccine. This time interval is needed for the body to adjust. If I take Temodar and receive the vaccine soon afterwards, with not even a day in between, who knows what will happen?

Papa sends an urgent email to Dr. Cooke. with copy to Dr. Parsa, to reschedule the vaccine injection.

No reply.

Papa calls Dr. Cooke. "Can we move the vaccine injection by one week," he asks, "to avoid a conflict with the Temodar?"

Dr. Cooke, without even acknowledging Papa's concern, repeats her instruction to abide by the pre-approved protocol for the heat shock vaccine. "Mino's vaccine injection has been scheduled for October twenty-sixth. Since Mino didn't take the chemo on time, he must take it now."

"You know perfectly well why Mino was unable to take the chemo on time. It was because of the grand mal seizure."

"I can't change the vaccine protocol. Mino must abide by the protocol. If not, we may have to declare him ineligible to continue in the clinical trial."

Papa gets visibly irritated. He requests that the case be presented to the UCSF Tumor Board at its meeting this week.

"Oh," Dr. Cooke says, "I'll be out of town by then...can't present it to our Tumor Board."

"Where will you be?"

"I'm flying to Florence for vacation."

Papa hangs up. He tries to reach the Head of the Neurosurgery at UCSF but can't get through.

I discuss with my parents what to do. If Dr. Cooke is unwilling to reschedule the vaccine, should we consider skipping the Temodar, which I take at home, and get the vaccine at UCSF as scheduled. In this case the conflict between chemo and vaccine could be avoided.

However, we have no way of knowing the risk if I skip the chemo. Asking Dr. Cooke is useless. Dr. Parsa is unavailable. The Head of the Neurosurgery can't be reached.

In the end, we don't have the courage to skip the chemo. If I don't comply with Dr. Cooke's order, UCSF may throw me out of the clinical trial. I know from Dr. Parsa that there were twenty-eight doses of the vaccine. Until now, I've received three booster injections at the beginning and a total of five monthly doses. Therefore, twenty left. What if UCSF declares me ineligible to continue the clinical trial, as Dr. Cooke darkly implied?

Dutifully, I take the chemo for five days and then off to UCSF on the following day, October twenty-sixth. The side effects of the chemo are always strongest on days four and five and remain strong up to day eight. As I walk slowly from the elevator to the vaccine injection room, my lower back pain is so intense that I fear fainting. Papa and Saya hold me from both sides. In an attempt to get a last-minute reprieve for my injection, Papa tries to see the department head but can't get beyond the secretary. "Sorry," the secretary tells him. "You should have made an appointment."

"But it's an emergency."

"Why don't you go to the Emergency Room?" comes the snappy answer.

Reluctantly, I submit to the vaccine injection. Then we drive home.

Less than twelve hours after the injection, a most violent reaction sets in, the feared collision between chemo and vaccine. The worst pain I've ever experienced, ten times worse than my dislocated L4. I scream at the top of my lungs. Saya and Papa rush to my bedside. The pain feels like someone is wielding a rusty razor blade to shred the muscles of my left arm, across my abdomen, and down the left leg. It feels as if someone is firing a nail gun into my body. I can barely breathe.

Not that I have never experienced pain before. I went through a lot as a child and a teenager. As a seven-year-old, I was in a bicycle accident on a rough trail, which resulted in about thirty small bits of gravel imbedded in my forehead. I still have a few left in my forehead. They work their way out once in a while. When I was thirteen, I suffered second- and third-degree burns when a flame shot out of a stove that I was stoking and hit my chest. And there was a dislocated shoulder, some ripped tendons, and a few broken ribs during a skiing accident in the Swiss Alps during my ETH years. But nothing in my memory compares to the torture I'm feeling now over the entire left side of my body.

Saya tries to help me with pain killers. No effect. My misery persists through the day and through the night, into the next day and the day after that, without a break. I'm unable to drink, unable to swallow food. I'm unable to sleep. I'm losing a lot of weight. But it's the pain that wears me out— the relentless pain—more than anything else. UCSF doesn't respond to Papa's urgent calls. All that he hears from them is that we should wait until Dr. Cooke comes back from her vacation.

From my analysis of the biochemistry of Temodar and the action of the vaccine in my body I had predicted that a violent collision between the chemo and the vaccine could occur. I was right. Our plea to delay the vaccine injection by at least a week had been countered with an unveiled threat that, if I didn't stick to the protocol, I might be thrown out of the clinical trial altogether.

While my traumatic pain continues, now already for four weeks, we still don't get any help from UCSF, not even a response to our repeated requests. Nothing. Nada. Just silence. Nobody returns Papa's phone calls or my calls.

Finally, in late November we are allowed to set up a meeting with Dr. Cooke. As usual, she is elegantly dressed under her white doctor's coat. She brags about her vacation in Italy and how splendid Florence is. She speaks with a gentle voice and smiles almost all the time. She performs her routine test of my peripheral vision by snapping her fingers to the left and right of my head, asking me to nod if I can see them. My pain is of no concern to her. On the contrary, she refuses to hear me out. It is quite clear that she is bent on letting me and anybody else know that she is in no way responsible for the messed-up chemo and vaccine schedule and the hellish pain in my body. Her only solution is to prescribe strong opiates like hydrocodone and tramadol—ugly drugs with nasty side effects like vomiting, constipation, exhaustion, and constant dizziness. She makes no effort to refer me to any pain specialists.

Several friends and colleagues at NASA urge me and my parents to sue the UCSF Neurology Department and Dr. Michelle Cooke for malpractice. They give us names of powerful attorneys who specialize in the malpractice field. However, we have no time and no energy for a lawsuit that would surely drag on indefinitely. Our fight is against my brain tumor. Our fight is for my life.

The only action Papa takes is to file a request with the Head of the Neurology Department to transfer me from Dr. Michelle Cooke to another member of the UCSF team, hopefully better qualified and more considerate.

In response, the UCSF Neurology Department sets up a meeting for us, but when we arrive, we are not met by members of the Review Panel, as we had expected, nor by the Head of the Neurology Department, but by a bulky, stately woman whom I have never seen before. She gives us a lecture how professionally competent all members of the UCSF Neurology Department are, how accomplished and internationally acclaimed. Top in the USA. and top in the world. This big woman has an oppressive way of talking. She dismisses Papa's comments and cuts him short. She points out that it is a privilege to have been accepted into the heat shock vaccine clinical trial, which UCSF pioneered; a *privilege*, she emphasizes again. Her whole demeanor reminds me of Cerberus in Greek mythology, the three-headed dog that guards the gate to the Underworld.

Help comes from Dr. Rose Lai, my neuro-oncologist at the Columbia University Medical Center in New York, who has diagnosed my pain across the continent as Reflex Sympathetic Dystrophy, also called Complex Regional Pain Syndrome, a condition that affects the sympathetic peripheral nervous system. Rose tells us that, of course, UCSF has an in-house pain clinic—even a pretty renowned one.

Suddenly I understand why Dr. Cooke and the Cerberus did not tell us about UCSF's in-house pain clinic. They don't want their colleagues at their own Pain Clinic to know that my pain is the result of their total incompetence and arrogance.

Rose refers me to the Stanford Pain Clinic and its Director Dr. Sean Mackey.

From late December onward, twelve weeks after UCSF plunged me into this torrent of pain, I'm receiving treatments at the Stanford Pain Clinic. They involve applying ganglion blocks to the nerves in my spine at the level of the cervical vertebrae in my neck and at the level of the damaged L4 in my lower back. Stanford addresses both the symptoms and the underlying causes of my pain in a daring way.

After the procedure has been performed on me for the first time, under local anesthesia, I become curious how the doctors do it. At home, I'm reading more about ganglions and learn that they provide the junction between the autonomic nerves of the central nervous system, and the peripheral nerves. A ganglion block is a highly specialized surgical procedure, during which a physician injects a pharmaceutical concoction, a ganglion blocker, into my spine. Directly into my spine, using thick needles up to twenty centimeters in length. To guide these needles, he uses tools such as ultrasound or X-rays, or just old-fashioned deep knowledge of the human spine.

The second time I receive the treatment, I tell the team in the surgery room that I don't want any sedatives. Considering the pain I've been through since my grand mal seizure three months ago, a little poke with a needle should not throw me off, even if it goes into my spine. The anesthetist looks puzzled, but I assure him that it's okay.

Now I watch the procedure on the computer screen and see how the needle is inserted into my neck and into my spine. I feel the needle going down into it, very deep. Then, with a dramatic twist, as the content of the syringe is emptied into my spine, the first ganglion block kicks in. Minutes later, a second needle is inserted into my lower spine. Again, with a dramatic twist, as the content of the syringe is emptied into my spine, the second ganglion block kicks in, taking away the pain, as if someone has turned off a switch.

In the beginning, the effect of the ganglion blocks tends to wear off after a few days, but repeated injections lead to longer relief times, up to three weeks.

Thus, with the help I'm receiving at the Stanford Pain Clinic and the continuation of my intense physical therapy regimen, my pain becomes tolerable. Occasionally it still suddenly flares up, but I can treat it with non-opiate painkillers

that have no bad side effects. In addition, Fred Dong's acupuncture helps me enormously. Plus I'm doing meditation.

Before the seizure, I could hike on any bumpy trail in Big
Basin among the redwood trees, up and down, for eight to
ten hours. I was doing well. Now I have to rely on my cane
everywhere I go, and pain is never far away. I can barely walk
more than half an hour a day, no matter how hard I try. I'm
back where I was one year ago.

REDWOOD GIANTS

I receive emails almost every day from friends and colleagues
at NASA as well as at other federal agencies donating their
leave time for me under the Voluntary Leave Transfer Program. They do it so that I can continue fighting my GBM. I
want to express my deep gratitude to friends and colleagues
who have been supporting me through this ordeal.

I know all this has been hard, physically and emotionally,
on my parents, though Saya continues to be cheerful and to
sing as before. Papa, who never sings, continues to look serious. Having reached their fiftieth wedding anniversary this
year, they are not the youngest anymore. Once in a while I
detect helplessness in their eyes, though both want to look
optimistic, even defiant. I've tried not to burden them, though
I could have screamed all day and all night.

Now I endure the pain whenever it flares up. I close my
eyes and try to breathe slowly. Something strange happens
to me, as if I were transported into a different state of reality. My breathing deepens. I'm in Big Basin, surrounded by
majestic redwood trees. I approach them, watching spots of
sunlight here and there. I touch their thick bark, listen to
their voices. I'm overwhelmed by a great inner commotion.
I feel pulled down, down to the ground, irresistibly, pulled
to my knees, touching the fallen redwood needles with the
palms of my hands.

It's one of those magic moments when all senses come together. I see the sunbeams swirling through the branches. I feel the earth. I hear the sounds. I smell the sweetness of the forest floor. I feel the energy streaming out of the redwood trees, streaming into me. An explosion of the senses, just for a flash, just for a moment; and as suddenly as it has come, the vision passes away. Then I realize that my mind has become utterly still. I inhabit a body without pain. I enter eternity, if only for that instant. The colors fade. The sounds are muted. I reenter reality, but the pain is gone, carried away by the vision.

At home, the cherry blossoms are all gone, but there are so many other flowers in the garden. The three apple trees are there in full bloom, with luminous white flowers tinged in pink. They look like a chorus from the Baroque era. The celestial bamboo bush in front of my window shines in the sun, with pastel greens and orange red with tiny round red berries. I'm grateful that I have started to rediscover all that exists around me without bone-crushing pain. I've also start-ed to eat again what Saya cooks for me, without throwing up. My mind is filled with many projects that are propelling to come out.

It's May and I go with Papa to the Stanford Nanotechnol-ogy Center for a workshop. One of my projects at NASA was to inject iron oxide nanoparticles, laden with drugs, into the blood stream and direct them with external magnetic fields to any spot in the brain where there is a tumor.

The invaluable advantage of such a targeted delivery tech-nique is that there will be no need to poison the whole body with the chemotherapeutic drugs. The big open question is how to overcome the blood-brain-barrier, BBB, which Nature has put up to protect the brain from all kinds of other dangers, in particular, from nasty bacteria. However, I'm sure that we will be able to find a solution to the problem with the BBB.

The iron oxide nanoparticles are my idea; pretty old by now, and others have adapted it too. The idea came to me before the diagnosis of my own GBM. Manipulating iron oxide nanoparticles with external magnetic fields is possible, because they are superparamagnetic. If we can attach the proper drugs to them, we can use them as carriers to deliver the drugs to where we want them to go. This will alleviate the need to flood the entire body with the anti-cancer drugs that cause many bad side effects.

I am thinking of that poor woman next-door in the UCSF Neurosurgery Acute Care, who was meowing all night. The surgery had damaged her speech center. I imagine, if the doctors could have delivered drugs directly to her tumor, bypassing the need for surgery, she might still be able to speak.

It's quite a distance to walk from the parking lot on the Stanford campus to the newly built Nanotechnology Center, but I manage to stroll along with Papa, securing my steps with my cane. After we arrive at the building and take the elevator to the second floor where the workshop is held, I quickly forget my exhaustion.

I run into friends from the nanofield and at once feel at home, scientifically. I sit through many of the talks and participate in some of the discussions. I'm able to renew some old connections and make plans for the future. One student asks me if he could come to work with me at NASA over the summer, and a second and a third line up behind him.

In the afternoon, however, I become so tired that I have to rest. The workshop organizers find a room where there is a couch for me to lie down. According to Papa, who sat on a chair in the corner, I immediately fell asleep.

By the time I wake up, the workshop has ended for the day, but there are still many participants standing around in the hallway chatting and sipping from plastic cups filled with ice water.

Next on my program, for June, is the International Brain Mapping Conference in San Francisco. I'm on the scientific program committee. I have organized one of the conference sessions, the one on Nano-Bio-Electronics. As session chairman, I can select the speakers.

One of my invitees is Dr. Sean Mackey, the director of the Stanford Pain Clinic. I'm interested in hearing what he has to say about the use of functional MRI to investigate neural processes during cognitive modulation of pain. Indeed, he reports how moderately painful heat can be used to test healthy people and to follow what happens in their brains when they are asked to fight the pain with cognitive control. Very interesting stuff. I wish Dr. Mackey would have been able to put me into his fast MRI machine during one of my own pain attacks.

Whenever I see Dr. Mackey at the Stanford Pain Clinic, I'm impressed by his eloquence. He is a very fast speaker, holding a small voice recorder in his hand. Just a couple of hours later, on the same day, when I open my computer at home, he has sent me a transcript of what he dictated into his voice recorder during our meeting. I marvel at the efficiency with which Dr. Mackey uses the voice recognition software to create a record of what happens during the day.

Another of my invitees is Papa—believe it or not. I have to be concerned that inviting him looks a bit like nepotism. But—what the heck—what Papa has been working on recently fits well into the theme of this brain conference. Actually, it's quite fundamental and a good example of how an interdisciplinary approach can bring out novel and unexpected insights.

For some time now, Papa has become known as the earthquake guy, but this is not the whole story. What he tries to do is to solve the physics of the processes that happen deep inside the Earth before major earthquakes. His approach is based on work we have published together over the years in top scientific journals. Now he is pushing ahead with some pretty nifty new ideas by taking his work to the next level. He tries to understand why animals have that remarkable ability to sense the approach of large earthquakes days before they strike. As part of this work it has become clear that humans possess the same ability. Importantly, this ability resides in the brain. That's why I invited him, nepotism or not.

Papa has shown me medical records from Canada that indicate an increase in the number of patients seeking help for migraine during the days leading up to moderate earthquakes. Medical records from a remote region in Russia's southern Siberia, where a very large earthquake hit in 2003, show that starting about ten days before the seismic disaster, the number of people seeking medical emergency help increased almost threefold. In nearly all cases, the people came to the ER because of neurological disorders such as hypertension, vegetative-vascular dystonia, and epilepsy, while the cases of acute respiratory infection and gastroenteric diseases increased only after the earthquake.

These observations have led Papa to investigate how the human brain reacts to subtle changes in the natural environment before earthquakes. He thinks he has to focus on the changes in the ultralow frequency radiation field, which includes the Schumann Resonances, very weak but sharp frequency bands created by electromagnetic standing waves. These waves travel around the globe, reflected off the Earth and the ionosphere.

Maybe there is something in the brain that is influenced by or couples to the Schumann Resonances. But what?

I told Papa about Viktor Stolc in my group at NASA Ames, who is working on the molecular motors called ATP synthase embedded by the billions in the membranes of all our cells. They are the energy engines of our bodies, and the brain has particularly many of them. The ATP synthase motors seem to rotate in synchrony at a frequency surprisingly close to the Schumann Resonances. Papa should work with Viktor to push this research ahead.

As the chairman, I'm sitting on the stage, off to one side. While the speakers present their papers one after the other, I look around to observe how the audience is reacting. Though the lights are dimmed, and the rows of seats are illuminated only by the reflection from the projection screen, I begin to recognize faces in the audience. I'm relieved to see that nobody is sleeping or looks bored. Nearly all seem to follow the talks with interest. Papa sits in the third row, next to Dr. Mackey. Saya sits in the last row on the left, smiling and nodding at me.

I'm overwhelmed by the thought that in one year, there will be another conference focused on the same topic, with new interesting results and new questions. I'm looking forward to that conference.

———

Between my medical appointments at the Stanford Pain Clinic, El Camino Hospital, UCSF vaccines, and TCM treatments, I manage to attend the All-Hands meeting and the next Executive Council at NASA as well as the Singularity University Commencement. This all is giving me an idea where we are heading.

I learn that NASA Headquarters is warming up to my ideas of nanosatellites, finally. It's such a promising development. I'm not against the large, complex, multi-billion-dollar satellites that weigh a ton or more. They are great and can do important missions. But with nanosats, we can do more— much more.

I know some people start to grin when I approach them with my nanosat ideas. "Here come the Minosats," they say in a teasing tone, but at least they have started listening. I never claim that my nanosats would replace the big-dollar birds. I only present the fact that they can do many things at one-tenth or one-hundredth of the cost. After all, we have to achieve as much science and collect as many valuable data as we can with a NASA budget that has the unfortunate tendency to shrink rather than grow from year to year. I'm happy that Pete Worden is fully behind my vision.

Around the same time, I finish my job as the Executive Officer for the DARPA FAST program. I can report that it demonstrated a significant improvement in photovoltaic power generation, up to 60-fold, 130W/kg as compared to the measly 3W/kg on the ISS today. I finish co-writing Boeing's Final Report and the FAST Applications Study jointly with Boeing, the Aerospace Corporation, and the Air Force Research Laboratory. At least I can now type with my right hand without pain.

Papa didn't tell me about the upcoming deadline for the NASA Ames Center Innovation Fund. I'm really mad at him. Only one day left till the deadline. I type the proposal in one sitting. It's about this fundamentally new idea predicting that rocks become softer when stressed. Nifty stuff. About ten days later I learn that my proposal has been ranked first among nearly one hundred submissions.

I also establish a contact with the "Blue Brain" project at the Swiss Polytechnic Institute in Lausanne. Dr. Bob Bishop tells me that the project will start within a few years with an exaflop computer, simulating brain functions down to the synopsis level.

I feel invigorated by all these activities and begin to see how I can develop these projects at NASA Ames as well as in cooperation with other institutions.

And, yes, before I forget, I think Papa and I have found a likely explanation for those unipolar pulses that shoot out of the depth of the Earth before earthquakes, only a fraction of a second long and related to the build-up of tectonic stresses. We believe there must be a volume pulsation, an instability that develops when a volume of rocks down there tries to expand but the pressure from miles of rock lying on top prevent it. If tectonic forces continue to push, there will be a point when the stressed rock volume overcomes the confining pressure and suddenly expands. In this moment an electric current bursts out, creating a unipolar pulse. Key to this explanation is the degeneracy pressure, yet another quantum mechanical marvel, one of the strongest forces in nature that prevents stars from collapsing to a single point after they have run out of fuel to power their nuclear fusion reactions. Papa is impressed that knowledge I've absorbed during my years in cosmology can help explain something here on Earth, the unipolar pulses, that seemed so puzzlesome.

RELAPSE

Yesterday, on June thirtieth, 2011, I had a routine MRI scan at the 3 Tesla MRI machine at El Camino Hospital. Today my parents are taking me to UCSF to discuss the MRI with one of the neuro-oncologists, Dr. Nicholas Butowski. He is representing Dr. Michelle Cooke, who is again on vacation. This was my ninth MRI scan since the beginning of vaccination. All of the previous eight MRIs had produced very clear results. No changes that might suggest anything unusual. Nothing but spotless images. So we are set for a routine discussion of this latest MRI with Dr. Butowski.

The good doctor shows us on his computer screen my MRI from yesterday. Pointing at the right parietal lobe where the first and the second tumors were surgically removed, he alerts us to a dark region, 2.2 x 2.1 x 1.3 cm in size. "I don't know what this is," he says, "but it looks suspicious."

After the endlessly long Fourth of July weekend I call Dr. Sarah Nelson, a world expert in radiology and biological imaging at UCSF. She promises to get a research-grade MRI scheduled for me tomorrow at UCSF Mission Bay. Her MRI machine has superb capabilities, including perfusion and diffusion studies measuring the blood flow in the brain. It can also do spectroscopy of metabolites. Thus, Sarah can distinguish between tumor tissue, scar tissue, and treatment effects.

On the day after the MRI in Sarah Nelson's group, we drive again to San Francisco, this time to meet Andy Parsa. We wanted to ask him about the heat-shock vaccine. After all, he developed this vaccine.

Dr. Parsa has already looked at the latest scan and starts to talk about another surgery: "Would be really easy," he says. "Just open your skull and see what it is that shows up in the MRI."

I ask him why he is talking about surgery. Could it be that something might be wrong with the vaccine? I'm not yet half through the twenty-eight doses that have been prepared for me.

We learn that the UCSF team has its weekly tumor board meeting this afternoon. I call Sarah Nelson. She agrees to attend and to present my case. In the end, all members of the board agree that I don't have a cancer recurrence. There are so many possibilities. The shadowy region could be a treatment effect, due to the immune response of my body. It could be a delayed reaction dating back to the violent collision between chemo and my vaccine. It could be an effect due to the intense high-energy radiation I received at Stanford in March and April of last year, not to mention the many CT scans, eighteen in total, which also use X-rays.

We have to wait for the next MRI, scheduled for August fourth, 2011.

Back at home, I realize that during the past ten months, the pain has taken over my life, and fighting it has become a priority. Too weak and distracted by the pain, I've not kept up with battling the GBM. Before, before the seizure, fighting the GBM was my first priority.

I reopen the folder on my computer in which I had collected, prior to the seizure, information about the arsenal of treatment options, though I was trusting the heat-shock vaccine. I never even remotely considered the possibility of a tumor recurrence while I was receiving my monthly vaccine injections. Dr. Parsa's surprise suggestion of a new surgery seems so much out of place.

Anyhow, UCSF offers no other treatment except the heat-shock vaccine. Nobody ever hinted at the possibility that this vaccine could fail. Well, I trust it won't fail. Why should it, since we are not yet even half through my stock of vaccine doses?

Stanford neuro-oncologists have nothing other and nothing better to offer. Except for surgery, radiation, and chemo, they have nothing else. We visited them and talked with the head of the department, who looked at me with bewilderment. "You are astoundingly clear about your situation. I'm sorry," he said.

Saya learns from her friend Mr. Sugawara in Stockton, that there is a place in Japan that has been producing anti-cancer vaccines for nearly half a century and has used them to treat cancer patients. It's the Hasumi Foundation. Saya calls Tokyo and talks directly with Dr. Hasumi. He explains to her that, in order to develop a vaccine specific for me, he would need a sample of my tumor tissue.

Around the same time, Papa finds out that there are possibilities in Germany too, one in Berlin and the other near Göttingen. According to the website, Berlin offers a clinical trial with iron oxide nanoparticles to deliver chemotherapeutic drugs to any location in the brain. I'm happy that somebody has already picked up on the idea with iron oxide nanoparticles, which I had formulated before the GBM struck me. Probably the people in Berlin came up with the same idea independently. A big problem, as I see it, would be that introducing superparamagnetic iron oxide nanoparticles into the brain would make it impossible in the future to do an MRI. I have long been aware of this problem. Now I'm wondering whether the group in Berlin has been able to find a solution. Papa contacts them and learns that they are also struggling with the superparamagnetism and have not found a solution.

That's why we turn to the other treatment option in Germany, the one near Göttingen. It does not require my tumor tissue but is based on dendritic cells that can be separated out of my blood stream. The procedure was worked out by Dr. Nesselhut and his team in Duderstadt. It is based on the expectation that my dendritic cells would have developed antigens during their interaction with my GBM cells and that these antigens would be transferred to the T-cells of my immune system. Since the T-cells are the soldiers of the immune system, my body would be enabled to fight my GBM.

The Hasumi vaccine in Japan and the dendritic cell treatment in Germany are fully approved by the regulatory bodies, but not by the FDA in the USA.

The next MRI on August 4th will tell us where we stand. "Tumor recurrence or no recurrence?" is the question hanging over us day and night.

During the week before the MRI, we fly to Los Angeles because Papa is involved—together with the Southern California Earthquake Center—in organizing a workshop at the University of Southern California.

Since we are already in Los Angeles, we drive under the hot summer sun to the UCLA Medical Center to consult with their neuro-oncologists as to what kind of treatment for GBM they can offer, in case my heat-shock vaccine loses its effect. We are told that there is Avastin to fight recurrent GBM.

Early on Dr. Rose Lai, my neuro-oncologist from Columbia, mentioned Avastin, but she had also said I should never take it. Avastin has the reputation of being the drug of last resort. It tends to work for a while, but then scores of tumors show up in the brain, and that's the end.

I'm very tired. I have difficulties walking even short distances and Saya stays by my side all the time.

The earthquake workshop is supposed to provide—here I'm quoting from the pamphlet—"a collaborative environment for researchers from different fields to explore common earthquake forecasting issues."

Papa gives the opening talk. He sets the stage for a discussion of the fascinating physics that underlies a bewildering range of reported precursory earthquake signals. He describes how the different signals are linked through physics. He emphasizes the potential of NASA space technology when combined with ground-based sensor networks and how this can be used to detect high tectonic stresses deep below prior to major earthquakes. A fantastic speech, I think, with well-selected viewgraphs; it's one of the best presentations I've ever heard. At the end I was ready to applaud.

I look around to see how the audience reacts. I'm expecting to see many of these earthquake experts, mostly seismologists from the USGS and other organizations, leaning forward in their seats while following the presentation. However, barely a sign of interest. Some look at their smart phones or computer screens. A few have nodded off. No questions, no comments. Why are we here, if this is supposed to be workshop for an interdisciplinary discussion?

Finally, one of the USGS big shots raises his voice and dismisses everything Papa has presented. A wholesale rejection: "Everybody knows that earthquakes can't be predicted," he pontificates with a stentorian voice. "All that nonsense we just heard about pre-earthquake signals—it's just trash."

I get mad and come to Papa's rescue: "You lack the basic knowledge of physics...the physics underlying pre-earthquake signals."

Big Shot retorts: "I'm also a physicist, a geophysicist."

"That's exactly the problem," I throw back at him, but the session chairman interrupts me and calls the next speaker.

Back home, my MRI on August fourth is again not clear. However, there are two observations that make me weary. First, my left ankle has become very weak and it fails whenever I try to walk. This symptom started in Los Angeles, and Saya could barely hold me and keep me from falling. She screamed for help. A security officer came running to support me from the other side. Second, there has been an undeniable increase in the size of whatever was showing up as a shadow in the MRI. Plus there is more edema.

Two weeks later, a new MRI shows that I do have a recurring tumor. It has already grown to 4.2 cm in diameter. Has to come out. I agree on the spot to a new surgery, my third. Dr. Andy Parsa does it at UCSF. It takes five hours.

The post-op MRI is extremely clean, just a cavity—no edema, no recognizable residual tumor tissue. According to the pathologist's report, the tumor tissue turned out to be a combination of treatment effect and one of the most aggressive and difficult to treat forms of GBM.

Papa sends a tissue sample to the Hasumi Foundation in Tokyo.

Five days after the surgery I choose among different rehabilitation facilities in the Bay Area a new place in Los Gatos, infinitely better than the Gloomy Light Rehab where I did time last year. My fighting spirit is as strong as ever. To hell with this "most aggressive and difficult to treat form of GBM" pessimism. I'll fight this battle and I'll win.

During the next ten days I try to regain control of my

body—standing up, walking without feeling insecure, typing with the fingers of both hands.

One night, I can't fall asleep.

For the first time, doubts sneak into my mind: what if I can't defeat my GBM?

What if the GBM defeats me?

I'm scared. I feel desolate.

Time creeps by. It's two o'clock in the morning. I call home. Papa picks up.

"Of course, I'll come immediately."

It's a half-hour drive.

I'm waiting.

I hear Papa's voice outside my window. He has to argue with security at the entrance, but then he makes it to my room. My fear dissolves and my desolation fades. Papa tucks in my blanket, as he always did when I was little. He lies down on the bed next to mine and covers himself with a blanket he has brought. I drift into sleep.

MEDIEVAL TOWN

Our room takes up the whole top floor of an old hotel. By pulling the heavy curtains in front of the windows we can make the room very dark. The church bell strikes every hour, but we stop hearing it after a day or two. We don't hear any noises from the hotel; it's as if we were alone in our own world.

After the flight from San Francisco to Zürich and the long drive over the Autobahn to Duderstadt near Göttingen, I slept for two days and two nights. Same with Saya. "It's the jet lag," she says, "particularly because we were flying eastward."

The room under the roof is almost ideal. Saya sleeps next to me; and Papa at the farthest corner, so he can snore to his heart's content. "He doesn't seem to experience any jet lag,"

Saya says, "because he snores like a bear." Her causation isn't quite clear, but I understand what she means. Papa can sleep in an airplane from takeoff to landing, as if he were at home in his bed, and he snores no matter what. By snoring, he seems to keep his own rhythm without getting into a jet lag. This is my interpretation of Saya's theory.

In the morning, we take the elevator down to the atrium-like breakfast room with sunlight falling through the glass ceiling. We self-serve different kinds of ham, eggs, fruits, bread, juice, coffee, and tea. As I can't walk by myself, Saya and Papa bring everything I like to the table.

Nevertheless, I'm sad that I've been unable to regain my ability to stand up and walk with my cane, even though I worked on it relentlessly for more than ten days, morning till evening, at the Los Gatos Rehab facility, always with several physical therapists around. I overheard a doctor saying in a low voice to a therapist that the nerve center in my brain in charge of my left leg might have gotten damaged by the third surgery.

I'm now confined to a wheelchair. Papa pushes it with his strong arms. Saya can't push it. She's such a lightweight. The world looks different from a wheelchair. The perspective changes, when the eyes are at the level of a five-year-old. The sidewalk looks more crowded, the street narrower, and the buildings taller.

Duderstadt is a small town and quite medieval. Most houses are timber-framed. Their exposed wooden beams are painted dark red and filled in with masonry in stark white or earthen yellow. The only stone structure is the church. From midmorning till nightfall the whole center of town is for pedestrians only.

Once upon a time, perhaps until one hundred-fifty years ago, Duderstadt was surrounded by an earthen wall, a remnant from the Middle Ages. Now the wall is an elevated trail, wide enough for Papa to push my wheel chair comfortably, with Saya walking next to me. We go every evening over the trail circling around the old town. It takes about one hour to complete the circle. On one side of the trail are the steeply gabled old houses with flower gardens, and on the other side there once must have been fields and pastures with a river flowing between them. Now a city extends there with the supermarkets, offices, some factories, modern houses, and highways.

I have already submitted twice to the separation of my dendritic cells in Dr. Nesselhut's practice, a small private clinic in an old timber-framed house in the center of Duderstadt. A team of physicians are applying their immunotherapy to cancer patients. But there is no elevator. Two strong muscular men hired from a local moving company carry my wheelchair and me up a long, curved staircase to a bright room on the second floor. Here I lie on a bed for three to four hours, while my blood is circled through a separator to extract my dendritic cells.

I see Saya sitting by the large window. Behind her stand two gabled houses with wine-red beams. They remind me of the day when Mr. Schwarzfeld took me to a village near Cologne. He began drawing a farmhouse, also gabled, and I drew a small house surrounded by trees. On that long-ago day the color of the light was almost the same as the color now outside the clinic window.

I would like to take up drawing again. Yesterday, while Papa was pushing my wheelchair, I already drew pictures in my mind, all from the perspective of a five-year-old. As we crossed the street and passed by a shop or a flower garden, I imagined how I would put them on my drawing block.

Eventually I should be able to stand up again and walk. I should be able to hike in the redwood forest all day, till darkness sets in. My nanosat projects will prosper and bring in billions of dollars. I'll use the money to buy all old-growth redwood trees so that nobody can cut them down. I'll use the money to collect millions of tons of plastic trash and remove it from the world's oceans.

How come nobody ever informed us about the immunotherapy options based on dendritic cells that could be separated from my own blood stream? How come all those neurologists and neuro-oncologists never mentioned the possibility of immunotherapy to harness the power of the antigen that my own body can produce?

I don't know if I should be angry or sad that we are here in this private clinic in Duderstadt, so far away from home, so late in my fight against the GBM, just because all these smart neurologists and neuro-oncologists, whose careers are dedicated to combatting brain tumors, never said a word to me about dendritic cells.

Even Dr. Rose Lai of Columbia, whom I trusted as a well-informed and caring neuro-oncologist, did not bring up the possibility of dendritic cell therapy. Together with my parents I had a Skype with Rose after my third surgery, frantically looking for the next treatment option. Her only suggestion was: "Why don't you take Avastin?" I still remember her smile on the Skype screen. In the past, she had always adamantly opposed Avastin. She had planted into me fear and disdain of this "horrible drug" as she called it. Now she was recommending Avastin.

As soon as I started reading up on dendritic cell therapy, I learned that, in Europe, it had been used for years with full approval of various regulatory agencies. Dendritic cell therapy is recognized as having the power to fight cancers. Not so in the USA. Or not yet.

At UCSF, nobody ever mentioned the possibility that my heat-shock vaccine could one day fail—might even be sure to fail. The UCSF neuro-oncologists must have known this all along, in particular Dr. Michelle Cooke. I once asked her what could be done in the event that the heat-shock vaccine failed. Her reply: "Then we should sit down and talk about the next option."

Knowing full well that UCSF has no "next option" to offer, why did she imply that they did have an option? She showed no sense of urgency. She had no willingness even to consider the unforeseeable...and if the failure of the vaccine was foreseeable, isn't it almost criminal not to have a second plan?

Unfortunately, I was too preoccupied fighting the pain, for too many months. Dr. Cooke took me off my antiseizure medication and gave me instead the stimulant Ritalin, causing my grand mal seizure. She forced chemo and vaccine onto me, one after the other, without any safe distance in between them.

Any medically trained person with some rudimentary understanding of how the body reacts to chemo and to the vaccine must know how dangerous this is. I suspect she was fully aware, but going on vacation to Florence, Italy, was more important for her. She even made a barely concealed threat that I could be thrown out of the heat shock clinical trial altogether.

As I think about it, a very frightening suspicion creeps into my mind. What is a clinical trial after all? A project to collect statistical data on how long patients survive after having received a certain treatment. Nothing less and nothing more. It's a project to collect statistical information about the number of months patients survive who have received treatment relative to patients who have not. The goal of a clinical trial is not to heal or cure. It's to produce numbers—cold, hard numbers.

Those numbers are believable only if the participants in the trial do not deviate from the protocol. If they do not go out on their own in search of alternative ways or better ways to combat their illness. The doctors involved in a clinical trial have no interest in seeing their patients getting better by doing something in addition to the prescribed protocol—anything which may prolong their lives, maybe even heal their illness. Any such action would compromise the statistical reliability of the numbers, the number of months until death, that is the only allowed outcome of the trial. What is the number of months the patients survived that received the vaccine versus the number of months the patients survived without vaccine? If the first number is larger than the second, then the clinical trial was a success, and the doctors who conducted it become famous and can brag about it.

Now I begin to understand why we ran into such a dogged resistance when we tried to reschedule one vaccine injection by one week or ten days. The fact that I had suffered a grand mal seizure did not count. The injection protocol took priority in the minds of those at UCSF who were running the clinical trial—absolute priority regardless of the outcome. That's why Papa had been unable to reach the head of the department, when he made his last-ditch effort to delay the vaccine injection in order to avoid the collision with Temodar. "Sorry, he has been called off to another meeting." That's why the Cerberus barked at us with such determination. Her only intent was to intimidate us. That's also why everybody on the medical side was unwilling to share any information. They have, in fact, no interest in helping. They only want to get those clean numbers, how many months to the time of death.

At least, Papa found Duderstadt. Late in the game but nonetheless.

Now, as I lie here, my blood is passing through a separator to extract my dendritic cells. Dendritic cells are potent agents, but they are not present in adequate numbers. So, after harvesting my dendritic cells, Dr. Nesselhut and his team must amplify them. They need much higher numbers. The procedure is well established. Takes a week. I have a week off.

BAURAT GERBER STREET

We drive to Göttingen, my childhood town. Around this time of the year, forty-five years ago, it was often rainy and cold in this part of Germany, but we feel the effect of climate change. It's a sunny, balmy, beautiful autumn day in 2011.

The cobblestone pavement along the Baurat Gerber Street is still the same. The neatly kept three-story stone houses to the left and right, mostly condos, look unchanged. Even the fences and the bushes in the small front gardens seem untouched by the passing of time.

Papa pushes my wheelchair up the street. We are eager to see the house where we lived from 1963 to 1968, from the time I was one to the time I was six. As the slope steepens, Papa must push my wheelchair more forcefully. I look at the houses on the left side of the street.

"There's our house," I cry out, "see that green balcony?"

Yes, that's it.

I turn to Papa and remind him, pointing at the sidewalk where we are standing, "Here you spanked me. You spanked me real hard."

For a moment Papa looks confused, but then, "Oh, yes, Mino. Long, long ago. You were only two. We came out of the house and you dashed ahead and ran onto the street between two parked cars. Had that car coming down the street not been able to stop a hand's width from you...it could have killed you."

Papa puts his arm around my shoulder. "True, I spanked you real hard."

"Yes." I nod. "I still remember those squeaking tires."

"Why did you have to spank him?" Saya's voice is full of reproach. "You could have explained to him how dangerous it is to run into the street."

I look up to the green balcony on the second floor and imagine my face peering above the railing. Mino is wearing his white hat. He's observing. There comes a bald man. "Mama, look. Uncle has no hair." "Minochan, Minochan, say it in Japanese, so that this man doesn't understand. He may get cross with us." There comes another man, also bald, very bald, who walks across the street towards our house. "Mama, look! Anohito hagechabin da!" Wrong again. This time it's a Japanese professor coming to visit Mama. He waves at me and smiles.

Inside the house how many times did I help Mama cleaning dishes, hanging my own diapers on the clotheslines on the backside balcony, and how many times did I play hide and seek with her? Then the Urashima tale. Mama had to crawl over the tatami playing Mother Turtle who carries me on her back. I am the good boy Urashima. Mother Turtle dives with me deep into the blue ocean water, swimming to her coral palace. Colorful fish greet us.

We had no TV at that time, but a record player and many LP records. I remember listening to them and dancing on our tatami with a huge mirror covering one side of the room. I still see Grandpa's calligraphy on a scroll hanging on the wall.

When the weather was fine, Mama took me out and we walked into town. I asked her to be a witch, covering her head with a red scarf and riding on a broom. I held on to the broom and we sang together, "Faster, faster over the river, faster, faster over the mountain, faster, faster, over the clouds...high up...high up."

Baurat Gerber Street is still the same some forty-five years on, as if time had stood still.

Looking from my wheelchair to the green balcony and to the window, I say, "I would like to live my childhood all over again."

ZÜRICH

Twenty-four hours after I received the first injection of the dendritic cell vaccine, I notice a palpable improvement. I'm regaining some of my left hand's agility. I can command my left foot to move, better than before. We contact Iris, a physical therapist whose office is in a narrow house next to the only remaining gate tower in Duderstadt. Under her care within a few days, I'm able to stand up and walk with my cane at least a few steps, with Papa ready to hold me, if my left leg threatens to fail. I would like to continue with Iris, but she already has set up plans for some vacation.

We leave Duderstadt for Zürich, a grueling trip under my condition. In Zürich, we meet Bob Bishop, a Board member of the Blue Brain project at the Swiss Polytechnic Institute in Lausanne. He is spearheading in Geneva one of the most ambitious environmental projects to model the entire Earth. We discuss the great challenge confronting him, because so many facets of the Earth are interconnected and have to be treated as such—as a system—from the core to the ductile mantle and from there to the brittle crust. Continents shifting. The oceans and the atmospheric skin of our planet, taking into consideration the powerful influences of the sun and the moon, and even more distant bodies in the solar system.

Dr. Bishop is on his way to Australia to see his daughter living down under. He plans to see me in California next spring to continue our discussion.

It's a strange feeling to be back in Zürich, where I spent ten years of my life, busy years, full of happy memories. However, nearly all who had been dear to me at that time are no longer around. Jørg Olsen is gone, Alex Müller is in a retirement home, and Mr. and Mrs. Arnold are gone, too, from whom I subrented a room in the eastern part of town, close to the wooded hills where I walked so often. Professor Ackert, whose hospitality I enjoyed during my first two years in Zürich, is too sick to receive us. Jan Korvink, my buddy and closest friend during my time at ETH, has moved on to a professorship in Germany.

And Amelia, my only true love.... There she stands among the wild gentian flowers, azure blue, which cover the slopes of the hanging valley. I hear her melodious Swiss intonation and see the dimples in her cheeks when she smiles.

With so many changes, which upset my memories, Zürich looks different. The city gives me a different feeling now, though the church towers, the bridges, the lake, and the far-away mountains look the same.

Papa asks me if he should drive me up to the Hönggerberg, but I shake my head. Why should I go back to the Physics Institute where only an echo exists of bygone days?

I fondly recall the chocolate store on the Bahnhofstrasse, where they always had homemade praline. I ask Papa to take me there. It's still the same, Sprüngli. I'm not supposed to eat any sweets, because cancer thrives on glucose and sugar. For over two years, ever since my GBM was diagnosed, I have strictly adhered to the no-sweets rule. Saya had drastically cut any carbohydrates from my diet. Now, in the Sprüngli store, I break the rule and buy a bag full of most delicious pralines.

We are going to meet Didier Sornette for lunch. I had hoped to sit at the Au Premier on the second floor of the Zürich Hauptbahnhof. That's where I had always gone during my ETH years when I had a reason to celebrate—like having made it to the top of my class in Theoretical Physics or having successfully demonstrated the superconductivity of the gold-organic single crystals, which Dr. Hilti had grown for me at Ciba-Geigy. Even Jørg Olsen didn't believe that they would be superconductive and, sure enough, for months the experiment didn't work. It's only after I applied pressure that the superconductivity kicked in. However, building a vise to apply pressure at liquid helium temperatures was no easy matter. I rewarded myself with a full course dinner at the Au Premier.

Today the Au Premier is closed, and we have to meet with Didier at the less prestigious restaurant on the first floor.

Didier is a pure-bred physicist whom Papa has known for years. He had been Professor of Mathematical Geophysics at the University of California in Los Angeles. That's where he learned about the work Papa and I have been doing on pre-earthquake physics. Didier immediately grasped the essence, even though most of the mainstream geophysicists, and particularly the seismologists, are still not getting it today. Now Didier is a professor at the ETH, heading a large group analyzing risks in many areas, including the economic realm.

What was supposed to be a carefree lunch turned into a passionate three-hour long discussion, mostly between Didier and me, spanning a range of deep scientific issues. Didier's quick and sharp mind introduces the essence of quantum concept as applied to the brain. We have a rapid-fire exchange of ideas about how the brain works, how thoughts are formed and how decisions are made, how quantum uncertainty enters into the process, and how the brain uses the enormous complexity of its neural networks to reach decisions.

Saya is sitting at the lunch table looking at me with some kind of bewilderment and a hint of worries that I will overreach the limit of my endurance. To the contrary, the discussion with Didier carries me aloft and I don't feel fatigue. I realize that, in Didier, I've found a soulmate with whom I could spend long hours exploring the world of ideas and new concepts. When I'm well again, I will resume my discussion with him either in Zürich or in California.

During this trip to Germany and Switzerland, I've been wondering what Heimat is for me. Taken literally, this German word means home or hometown, or *kokyo* in Japanese. It means the place of birth and of upbringing. In my case, it could be Penn State. It could be Göttingen, or a bit of Chaperon Rouge in Crans-sur-Sierre in Switzerland. It could be Cologne or Zürich. However, in reality, Heimat doesn't mean any geographical place. More important is that Saya and Papa are near and that I would be surrounded by a few friends in the world of science and art with whom I can explore the Unknown.

WATERCOLORS

Now that I'm back in California, Pete Worden comes to visit me. We talk about so many things, particularly, how the NASA Center is doing. There have been changes in the upper management. Now Pete has more things on his plate than he already had before.

"Pete, don't worry. I'll get rid of this stupid tumor and come back to help you." I try to smile but have to fight back tears. As Pete leaves, he gives me a big hug, "See you, then," he says.

The dendritic cell vaccine seems to really be working, even though I have not yet regained the ability to walk even a few steps unsupported.

We have to fly to New York, all three of us, so that I can get my next injection of dendritic cell vaccine by Dr. Nesselhut, who flies in from Germany. All this only because the FDA forbids American doctors to do any such injections, even though the drugs to be administered have been approved in other advanced countries.

The flight was extraordinarily arduous, the airport overflowing with passengers and me on the wheelchair. Security, security, security, TSA agents particularly suspicious of my bag full of medicine and supplements. I had to get out of my wheelchair, which was taken apart to look for hidden weapons of mass destruction or a ton of narcotics, while I had to stand with Papa and Saya holding me from both sides to prevent me from falling. Same procedures in San Francisco and New York, each time half an hour at least to make sure that I'm not a terrorist in disguise.

Coming back to California I immediately have my next MRI. The region where the tumor had been still looks good, but now it's the radiologist who becomes concerned. The midline of my brain has shifted to the left more than it should. The reason is that internal pressure is building up in my right brain, obviously because the dendritic cell treatment has kicked in. It looks like the dendritic cells have learned to recognize my GBM and transferred this knowledge to my T cells, the soldiers of the immune system. My T cells seem to have started attacking the GBM massively.

That's good news, isn't it?

But there is a drawback. The brain tissue is swelling as a result of this war in my head. The swelling produces an additional edema. It pushes on the midline. This is dangerous. The doctors put me immediately back to a high dose decadron, intravenously, to reduce the swelling.

We all breathe a sigh of relief.

I want to draw again—perhaps in watercolors. I go on the web and order a collection of brushes and watercolors, and blocks of paper of the best quality I can find. We still have my old easel somewhere in our garage. I ask Papa to bring it to my bedroom so that I can look at it and imagine how it will be when I put a fresh white sheet of watercolor paper on it.

In my mind, I start mixing the colors on the palette. I see myself throwing a few strokes in crimson red on the paper, and shades of garnet and amethyst purples reflecting

the autumn colors of the trees that line the streets near our
house. I hear Scriabin in my mind, as I hear him often these
days. More strokes bursting into flames of light. The autumn
trees with the azure blue of the California sky gives me an
intense feeling of life.

I'm not yet done exploring.

SAYA SINGING

I sleep most of the time. I have no pain. I feel like floating. It's
getting more and more difficult to formulate words, though
I can hear everything that is said around me.

"We'll take good care of him." That's Dr. Dormady's
voice. He rushed to my bedside at home after I could no
longer drink and swallow food.

Somehow I have been brought to El Camino Hospital.
I barely remember how, but it was by ambulance. Without
sirens.

"We are high up on the sixth floor." Saya describes the
view from my room. "A large window on the East side overlooking the trees below. Far away the mountains."

I don't feel like opening my eyes. The darkness is soothing.

Over the course of the last few days, I have noticed that
my peripheral vision is shrinking. It's as if darkness were
slowly moving in from the edges of my field of view, almost
imperceptibly but without return.

My hospital bed is obviously surrounded by much medical equipment. It clicks and thumps incessantly, occasionally
there is a beep.

A nurse comes and takes my vitals. "Which college does
he go to?" I hear her asking Saya. Her voice sounds cheerful
and young.

"Oh, he is through college, years ago."

"But he looks so young. Reminds me of my boyfriend."

After the nurse is gone, Saya bends over me and presses

her face against my cheek. In a low voice, she starts to sing. She is singing *"Sho-sho—shojoji"*. I see myself dancing with her in our tatami room. We enter a temple garden in the full-moon night. Raccoons appear out of nowhere, dancing with us. I tap my drum *Pon poko pon no pon.*

I'm drifting back into sleep. I dream Papa picks me out of the stroller. Two tiny Easter ducklings are crossing the trail in front of us, tumbling all over each other. They are lost and don't know where to go. Papa puts one of them into my hands. Hearing the intense distress call from such a tiny body—chirps, twitters, and trills so loud, I feel like crying, too. We carry both ducklings to the nearest pond and put them into the water below the willow tree. They stop wailing and swim away. We have found them a home, and I wish them a mama.

"It's New Year, Minochan," Saya says. "New Year's Day, 2012. The sun is bright, but the air is chilly. The far-away mountains look almost blue." She promises to take me back to Big Basin, to the majestic redwood trees, as soon as we're out of here.

Yes, Redwoods, take me back...take me back. Among them, life is going at a pace that is measured in centuries.. They are giants for whom time has come to a crawl. They grow rapidly during their first fifty to one hundred years, then enter a juvenile phase, and finally reach adolescence at the age of about eight hundred years. They continue to grow to full adulthood for another thousand years, reaching their full height. Despite their enormous girth and size, the redwoods are gentle. There is no violence in them, no hastiness—only the never-ending cycle of life.

Everything is moving. Everything is flowing. Everything is evolving. All the time. The only constant is change. *Panta rhei.*

The redwoods have a calming effect on me, deeply

calming. When I'm among them, I feel unfazed by pretty much anything. It is as if the whole world has become a mirage.

I hear Dr. Dormady's voice. "Mino, your last MRI shows that the midline shift has eased. This is a very good sign."

I try to nod, but I'm not sure whether or not my muscles responded.

Since I have lost my ability to drink and swallow, and darkness has started to cover my eyes, I am yielding to the grim reality. I no longer believe that I can win my battle against the GBM. The tumor is still progressing in my brain, shutting off neurons after neurons.

I have fought for many months. Amidst extreme pain. I've fought for twenty-nine months altogether, nearly two and a half years—much longer than anyone predicted.

It is wonderful to hear Saya sing. She still believes I can win.

When the nurses come to change the linen of my bed, Papa holds my head still, while they roll me over and back. It's so comforting to feel Papa's hands, cool and strong.

Whispering close to my ear, Papa says, "Mino, I'll continue our work."

———————

Saya has stopped singing.

FINALE

Don't stand over there, mourning
Over there is not where you can find me
Though I did die, I am here
I am the rustle in the branches of the redwood trees
I am the sunbeam on the hummingbird's wing
I am the waterfall in the mountain
I am the silence before sunset over the ocean
I am the star that tells the story of the universe
I am here and will always be
As long as you remember